Factory Design for

Modular Homebuilding

Equipping the Modular Factory for Success

Michael A. Mullens

Constructability Press

Winter Park, FL

ISBN: 978-0-9833212-0-0

Library of Congress Control Number: 2011901701

Constructability Press
Winter Park, FL

This book is dedicated to the two women in my life – my mom Elaine and my wife Ellen – who taught me to love the pursuit of knowledge and who gave me the freedom to do it.

PREFACE

BOOK ORIGINS

In 1990 I accepted my first (and only) academic position, as a faculty member in the Department of Industrial Engineering and Management Systems at the University of Central Florida. My first research assignment was to explore the use of innovative manufacturing technologies in prefabricated homebuilding. I pursued this research topic throughout my academic career, and I continue to pursue it today, in my retirement, serving the prefabricated homebuilding industry as a consultant. Before entering academia, I served as an engineering consultant for more than a dozen years, designing highly automated manufacturing and distribution systems for leading manufacturers in aerospace, automotive, defense, electronics and healthcare. I can still remember my initial overconfidence and oversimplification as I naïvely attempted to interest prefabricated homebuilders in the same advanced manufacturing technologies. In retrospect, the prefabricated homebuilding industry is one of the most complex manufacturing environments that I have ever encountered. There are many complicating factors: diverse product offerings allowing partial or even complete customization; building systems and materials developed for site-built construction; numerous sub-systems (structural, electrical, plumbing, HVAC, data/communications, finish) that must be integrated in the product; strict regulatory codes that vary by locale; large heavy components; and numerous manufacturing activities involving fabrication, assembly and finishing.

Over time, members of the prefabricated homebuilding community enlisted as partners in our research. They guided research direction, reviewed results and graciously opened their factory doors, allowing the factories to become working laboratories. Research efforts were enabled by continuous funding over two decades from the U.S. Department of Energy's Building America Program, the U.S Department of Housing and Urban Development and the National Science Foundation. This book organizes and documents many of the resulting research findings.

FOCUS

The focus of this book is modular homebuilding, one of the most promising approaches to prefabricated homebuilding. It uses large, three-dimensional, factory-built modules to build a home. Each module includes the floor, walls and ceiling/roof with plumbing and electrical systems installed and interior and exterior finishes applied. The modular home is usually 85–95% complete when it leaves the factory. Utilizing modern industrial technology, modular homebuilding offers the promise of building a higher quality home, faster and at a lower cost. In practice, however, modular homebuilders and homebuyers have found this theoretical promise to be elusive. This book examines the theory and practice of modular homebuilding, identifying its strengths and weaknesses and introducing a structured engineering design approach for configuring a high performance modular factory. The approach can be used for the design and operation of a new greenfield factory, the redesign of an existing modular factory or the redesign of a general purpose facility for modular homebuilding. The book provides the first substantive discussion of factory design for modular homebuilding. My goal is to equip stakeholders inside and outside the industry – factory designers, product designers, operating managers, investors, and researchers – so they can design and operate high performance modular factories.

The key strengths of this book are an aggressive vision for a high performance modular factory and a structured engineering design approach for achieving it. Both are the result of integrating three disparate, but invaluable sources of knowledge:

- The practices and experience of current modular producers, many of whom are industry pioneers who helped create and develop the industry.
- Emerging best business practices, such as lean production and mass customization, that are transforming the industry.
- Current scientific research findings that provide insight about the industry.

This book is organized into five chapters:

- Chapter 1 – defines modular homebuilding, envisions its promise, and describes the present reality. Challenges that have eroded the success of modular homebuilding are identified.
- Chapter 2 – uses the key findings from Chapter 1 to develop a production strategy that guides factory design. Lean production and mass customization are prominent in the production strategy.
- Chapter 3 – describes the primary building elements used in modular homebuilding and the production processes used to produce them.
- Chapter 4 – introduces a six-step structured engineering design approach for configuring the modular factory.
- Chapter 5 – discusses the implementation of the modular factory.

The focus of this book is modular homebuilding in the U.S. All statistics are from the U.S. industry. It presumes wood frame construction, the predominant building system used by U.S. modular producers. However, the basic principles and methodology are robust and apply to prefabricated homebuilders throughout the world. For example, they can be used by modular producers using alternative building systems such as light gauge steel framing, structural insulated panels (SIPs) or other composite materials. They can be used by HUD Code home manufacturers in the U.S. They can also be used by producers of smaller-scale prefabricated components, such as bathroom modules or wall panels. Finally, they can be used by prefabricated homebuilders throughout the world. It is important to emphasize that even though this book does not address prefabricated homebuilding outside the U.S., this does not mean that international prefabricated homebuilders are irrelevant. In fact, visits to leading prefabricated homebuilders in Japan, Sweden and the U.K. have demonstrated that, in many ways, they are more advanced than their U.S. counterparts. They are very large and well capitalized, allowing them to become the most technologically sophisticated housing providers in the world and offering the best examples of automotive-scale business practices in homebuilding. They use their size and capital as leverage for vertical integration – encompassing research and development, marketing and sales directly to the homebuyer, manufacturing, field installation, and long term customer service and follow-up. While their home designs are visually simpler than most U.S. designs, the design details are more complex to meet critical performance needs, such as energy efficiency in Sweden or earthquake and fire resistance in Japan. They are more likely to use advanced manufacturing technologies to enhance quality, increase productivity, and improve ergonomics. Several examples are particularly notable. Swedish factories emphasize ergonomics, utilizing production/material handling equipment and workplace design to minimize physical activity of workers and lifting heavy loads. This reduces worker fatigue and injuries, lowers worker turnover and allows homebuilders to retain older, highly skilled workers. In summary, factory workers are treated like factory workers, not construction workers. Japanese factories demonstrate this same emphasis on ergonomics, while placing additional emphasis on automation to improve quality and productivity. Prefabricated homebuilders in both Sweden and Japan place great emphasis on quality, and they are trusted by homebuyers to deliver excellent quality and service.

Given the many strengths of these international homebuilders, it is reasonable to ask why U.S. homebuilders should not simply clone their production strategies. Some insight can be gained from U.S. industry executives who shared their thoughts during plant visits to Sweden and Japan. They expressed great admiration for the achievements in quality and ergonomics. However, they were openly skeptical about the extensive capital investment required to attain what appeared to be only modest gains in production capacity and productivity – in other words, how can these staggering investments pay for themselves, particularly in the U.S. market with its wild and unpredictable demand cycles. Their skepticism was reinforced by recent events in the U.S. homebuilding industry, particularly the failure of the most advanced prefabricated housing factory in the U.S. [1] Pulte Homes, one of the biggest site-built homebuilders in the U.S., created Pulte Home Sciences (PHS) to prefabricate large-scale shell components for its homebuilding operations. Prefabricated components included pre-cast concrete foundation walls, floors built using open web steel trusses, SIPs for exterior structural walls and interior walls built using light-gauge steel framing. PHS used advanced CAD modeling to design, engineer and drive its highly automated production processes. The PHS factory had a capacity of 2,000 homes per year. The predominant reaction to Pulte's effort from both financial analysts and competitors was skepticism and can be summarized in the following quote from a Wall Street housing analyst [2]: "This is still a cyclical business, and the reason [builders] hire subs is so they can get rid of them during slower periods." In January 2007, three years after the PHS plant was opened, Pulte management announced the plant's closing [3]:

> *"Due to current market conditions, the plant was operating at about 25 percent capacity by the end of 2006 and on [Jan. 26] we reached the decision to close the facility .... Longer term, pre-manufacturing and transporting sizeable, heavy components to a geographic area limited to within 125 miles of the facility proved economically unviable, especially under current market conditions .... Overall, costs were higher and that was something that Pulte Homes could not pass on to home buyers."*

Consequently, this book is not about how to clone the production strategies of international homebuilders for use in the U.S. Instead, it focuses on how to add value for the homebuyer using the inherent advantages of prefabrication, in a way that is cost effective and sustainable in the U.S. These principles should be useful for all prefabricated homebuilders, even advanced homebuilders, anywhere in the world.

INTENDED READERS

This book is directed to several groups:
- Industry professionals that are responsible for owning, designing and operating a modular factory.
- Advanced undergraduate and graduate students that are studying residential construction, construction science, construction management, building technology or industrial engineering and who are enrolled in courses addressing prefabricated homebuilding. The book is most likely to be used as a supplemental text, but could be used as the primary text for a more tightly focused course.
- Faculty and students that are engaged in academic research involving prefabricated homebuilding.

REFERENCES USED IN THE PREFACE

1. Mullens, M., " Innovation in the U.S. Industrialized Housing Industry: A Tale of Two Strategies," *International Journal for Housing Science and Its Applications*. 32(3), 163-178, May 2008.
2. Caulfield, J., "Bold Maneuvers: Vertical Integration and Supply Chain Focus Take Shape as Pulte Looks to Double Capacity over the Next Three Years", *Big Builder*, July 1, 2004.
3. Burney, T., "Supply Side: Found Money", *Big Builder*, March 15, 2007.

ACKNOWLEDGEMENTS

My thinking on prefabricated homebuilding has been influenced by many people over the last 20 years. I want to thank each and every one of you. Although I cannot personally acknowledge everyone, I do want to personally thank each of you who have made special contributions. I want to start by thanking the sponsoring agencies who provided necessary research funding, without which little gets done in academia. The U.S. Department of Energy sponsored the Building America research program and its predecessor, the Energy Efficient Industrialized Housing research program. Mr. George James at DOE provided support and critical direction throughout these long term programs. The National Science Foundation sponsored a cooperative research project with the MIT School of Architecture that focused on open building systems and their potential impact on the modular factory. The U.S. Department of Housing and Urban Development sponsored a partnership with the Manufactured Housing Research Alliance (now the Systems Building Research Alliance) to introduce lean production to the industrialized housing industry nationwide. The New York State Energy Research and Development Authority sponsored a similar partnership to introduce lean production to industrialized housing producers in New York. This effort also extended the use of lean beyond the factory to include sales and installation on the building site.

Next I want to thank my primary research partners in these programs. These include Dr. Subrato Chandra at the Florida Solar Energy Center who directed the Building America Industrialized Housing Partnership, Mark Kelley who directed the Building America Hickory Consortium, and Emanuel Levy and Jordan Dentz at the Systems Building Research Alliance. Rick Boyd of Clayton Homes, who also serves as Chair of the Systems Built Research Alliance, should also be recognized for his unique role in opening the industry to lean improvement.

I also want to thank my fellow researchers, both faculty and students, at the University of Central Florida Housing Constructability Lab. UCF faculty included Dr. Bill Swart, Dr.Bob Hoekstra, Dr. Bob Armacost, and Dr. Ahmad Elshennawy. Although there have been too many students to thank individually, I want to especially thank our Ph.D. students who have graduated and are now teaching and performing research in prefabricated homebuilding. These include Dr. Isabelina Nahmens, Assistant Professor in the Department of Construction Management & Industrial Engineering at Louisiana State University and Dr. Mohammed Arif, Senior Lecturer/Programme Director MSc Advanced Manufacturing in Construction, at University of Salford in the U.K.

I also want to thank my colleagues at the National Consortium of Housing Research Centers who are dedicated to strengthening U.S. housing research. Dr. Matt Syal at Michigan State University and Dr. Mark Hastak at Purdue have contributed much to my research.

Special thanks are due to my colleagues who reviewed this book and made countless suggestions. They include Dr. Randy Cantrell at the National Association of Home Builders Research Center, Jordan Dentz at the Systems Building Research Alliance and Dr. Isabelina Nahmens at LSU. The book is better because of their efforts.

Finally, I want to thank the many women and men of the industrialized housing industry who worked closely with our research team over many years. We learned much from them. I particularly want to thank the industry leaders and their firms who worked closely with us in the Quality Modular Building Task Force. They helped to set our annual research agenda and reviewed our findings. In the end, they implemented the results that they viewed as valuable. These include Neil Sayers of All American Homes, Bret Berneche of Cardinal Homes, Doug Basnett of Epoch Homes, Ed Langley of Excel Homes, Noel Ward of Nationwide Homes and Roger Lyons of Penn-Lyon Homes.

To assist the reader who may be unfamiliar with modular homebuilding, the book is highly illustrated. I want to thank Alouette Homes, Cardinal Homes, Epoch Homes, Excel Homes, Nationwide Homes and Titan Homes for graciously allowing me to use photos of their products and factories in this book. Most of these photos were from a vast library taken over two decades of research. Consequently (and regretfully), I did not obtain permission from workers to use their images. Even though I have been forced to distort their images to prevent identification and protect their privacy, I want to thank them for their role in this effort.

TABLE OF CONTENTS

CHAPTER 1   INTRODUCTION TO MODULAR HOMEBUILDING: PROMISE, REALITY AND CHALLENGES ... 1

1.1 WHAT IS A MODULAR HOME? ... 1
1.2 WHAT IS MODULAR HOMEBUILDING? ... 1
1.3 THE PROMISE OF MODULAR HOMEBUILDING ... 6
1.4 THE REALITY OF MODULAR HOMEBUILDING ... 11
  1.4.1 Encouraging signs ... 11
  1.4.2 Modular production statistics and market share ... 14
  1.4.3 Remaining challenges ... 15
1.5 ATTAINING THE PROMISE ... 25
1.6 REFERENCES ... 26

CHAPTER 2   A PRODUCTION STRATEGY FOR MODULAR HOMEBUILDING ... 29
2.1 MODULAR PRODUCTION STRATEGY: OVERVIEW ... 29
2.2 PRODUCTION STRATEGY: KEY FOCUS AREAS FOR FACTORY DESIGN ... 31
  2.2.1 Capacity management ... 31
  2.2.2 Lean production ... 32
  2.2.3 Mass customization ... 39
2.3 REFERENCES ... 43

CHAPTER 3   MODULAR HOMEBUILDING COMPONENTS AND PRODUCTION PROCESSES ... 45

3.1 THE MODULE ... 45
  3.1.1 Limitations on Module Size ... 45
3.2 COMPONENTS AND PROCESSES ... 48
  3.2.1 Cut framing components (mill) ... 48
  3.2.2 Build floor ... 50
  3.2.3 Build window and door opening sub-assemblies ... 51
  3.2.4 Build partition walls ... 52
  3.2.5 Build side wall ... 54
  3.2.6 Build end walls ... 56
  3.2.7 Build marriage wall ... 57
  3.2.8 Set partition walls ... 57
  3.2.9 Set exterior and marriage walls ... 59
  3.2.10 Install rough electric in walls ... 59
  3.2.11 Build plumbing subassemblies ... 61
  3.2.12 Install rough plumbing in walls and tubs ... 62
  3.2.13 Build subassemblies for roof ... 63

      3.2.14 Build roof/ceiling   65
      3.2.15 Set roof   67
      3.2.16 Install rough electric in roof   69
      3.2.17 Install rough plumbing in roof   70
      3.2.18 Insulate roof   71
      3.2.19 Sheath and install subassemblies for roof   72
      3.2.20 Shingle roof   74
      3.2.21 Install fascia and soffit   76
      3.2.22 Prep and drop roof and wrap for shipment   76
      3.2.23 Insulate walls   77
      3.2.24 Sheath walls   79
      3.2.25 Install windows and exterior doors   79
      3.2.26 Install siding and trim   80
      3.2.27 Hang drywall on walls   81
      3.2.28 Tape and mud drywall   81
      3.2.29 Sand and paint   82
      3.2.30 Install cabinets and vanities   82
      3.2.31 Fabricate and install kitchen countertops   83
      3.2.32 Build finish plumbing subassemblies   83
      3.2.33 Install finish plumbing   84
      3.2.34 Install finish electric   84
      3.2.35 Build interior door subassemblies   84
      3.2.36 Install interior doors   85
      3.2.37 Install molding   85
      3.2.38 Install miscellaneous finish items   86
      3.2.39 Install flooring   86
      3.2.40 Load shiploose materials   87
      3.2.41 Factory touch-up   87
      3.2.42 Install plumbing in floor   88
      3.2.43 Load module on carrier   89
      3.2.44 Final wrap and prep for shipment   91
      3.2.45 Build major shiploose subassemblies   92
   3.3 REFERENCES   93

CHAPTER 4   DESIGNING THE MODULAR FACTORY   95

   4.1 DESIGN PRODUCT AND DEVELOP PRODUCT
       ARCHITECTURE   96
   4.2 DEVELOP SALES PLAN   97
   4.3 DEVELOP CAPACITY PLAN   97
   4.4 IDENTIFY AND DESCRIBE VALUE-ADDED   PROCESSES   98
      4.4.1 Estimating labor requirements and cycle time   99
      4.4.2 Estimating Cycle Time Variation   102
   4.5 MAP THE VALUE STREAM   103
      4.5.1 The VSM   103
      4.5.2 Using the VSM   110

4.6 DESIGN THE PRODUCTION LINE 144
4.6.1 Sidesaddle line layouts 146
4.6.2 Shotgun line layouts 155
4.6.3 Build-in-place 165
4.6.4 Roof access 167
4.6.5 Material handling for modules 173
4.6.6 Material handling for major subassemblies 177
4.7 DESIGN THE FACTORY FACILITY 193
4.8 DESIGN THE WORKPLACE 201
4.8.1 Case study 202
4.8.2 Handling building materials in the workplace 208
4.8.3 Secondary storage for building materials 218
4.9 REFINING THE DESIGN 220
4.10 REFERENCES 221

CHAPTER 5 IMPLEMENTATION: REALIZING THE PROMISE 223

5.1 PLANNING 223
5.2 PROCUREMENT 224
5.3 FIT-OUT 227
5.4 STARTUP 227
5.5 ONGOING OPERATION 228
5.1.1 Optimum Value Engineering for Framing 228
5.1.2 Supersize Building Materials 230
5.1.3 Real-time Shop Floor Information System 230
5.6 REFERENCES 231

INDEX 233

CHAPTER 1
INTRODUCTION TO MODULAR HOMEBUILDING:
PROMISE, REALITY AND CHALLENGES

Chapter 1 introduces the reader to modular homebuilding. It defines modular homebuilding, envisions its promise, and describes the present reality. Challenges that have eroded the promise of modular homebuilding are identified so that they can be addressed in the design of the modular factory.

1.1 WHAT IS A MODULAR HOME?

A modular home is a home produced using factory-built modules. A modular home looks and performs like a conventional site-built home (Figures 1.1–1.2). One can argue that the term "modular" home ceases to have meaning after construction is complete. Modular home design ranges from a standard model to a unique and one-of-a-kind design. Most often, a modular home design is based on a standard model and customized to meet the unique needs of the homebuyer. Modular home design options are virtually unlimited, offering a wide variety of architectural styles, sizes, levels of finish and cost. A modular home is built using the same construction materials and construction details as a conventional site-built home, with minor changes to accommodate shipping and installation. A modular home complies with state and local building codes similar or identical to those that apply to conventional site-built homes. Most modular homes are located in states that have adopted a preemptive state modular building code that is based on a major model building code (International Building Code). These preemptive state modular building codes cannot be amended locally, assuring uniform statewide access to consistently well-designed and well-built modular homes. Almost all modular homes are produced using wood frame construction. However, light-gauge steel framing, structural insulated panels (SIPs) and pre-cast concrete panels are also being used to produce modular homes. Carlson [1] and Gianino [2] discuss modular homes in greater depth.

1.2 WHAT IS MODULAR HOMEBUILDING?

Modular homebuilding is an innovative approach to homebuilding that uses large, three-dimensional, factory-built prefabricated modules to build a home. Each module includes the floor, walls and ceiling/roof with plumbing and electrical systems installed and interior and exterior finishes applied. The homebuilder orders the modules that are needed to build a home from the factory, working closely with the factory sales and engineering staff to fully specify the home design. After the order is placed, the factory orders unique components from suppliers and places the home on the production schedule. Scheduling depends on the homebuilder's schedule, the factory production queue, and the expected delivery dates for unique components ordered from suppliers. Once production starts, the factory production cycle is typically three to six days. After completing factory production, modules are loaded onto carriers and transported to the home site. The modular home is usually 85–95% complete when it leaves the factory.

Figure 1.1 Modular homes

Figure 1.2 Modular homes

At the home site, the set crew uses a construction crane to lift modules off the carriers and set them on a permanent foundation. The set crew then attaches the modules to the foundation and to each other, raises any folding roof elements and seals the home so that it is weather-tight. The entire set process is typically completed in one to two days. Mullens [3] describes the modular set process in greater detail.

After the home is set, it is finished by the homebuilder. Although the remaining finish activities are much simpler than conventional site-built construction, they are not trivial. In addition to the site and foundation work that is completed before module delivery, the homebuilder must finish the home and add amenities such as a porch, deck, garage and landscaping. Finish activities include: finish interior and exterior marriage joints (between modules) on the floors, walls, and ceilings; connect utilities; add heating, ventilation and air conditioning (HVAC) systems; and install fragile specialty items such as ceramic tile. The finish process typically requires four to twelve weeks. Mullens and Kelley [4] describe the modular finish process in greater detail. A pictorial summary of the modular homebuilding process is shown in Figure 1.3.

Modular homebuilding differs from conventional site-built construction, in which all of the elemental building components (such as dimensional lumber) are delivered to the home site where they are fabricated, assembled and finished. It should be noted that conventional site builders are increasingly turning to factory-built components (such as roof trusses) to increase quality, reduce costs and speed construction.

Modular homebuilding is one of several approaches used in prefabricated homebuilding. Other approaches include panelized, HUD Code and whole-house homebuilding. Panelization is the most common approach to prefabricated homebuilding. Panelized homebuilding uses two-dimensional, factory-produced panels to build a home. Almost all panelized homebuilding systems use open wood frame wall panels, consisting of dimensional lumber framing and exterior structural sheathing for exterior panels. When suitable, factory-produced floor and roof panels are also used. A few panelized homebuilders use closed wood frame panels, in which the factory installs plumbing, wiring, and insulation in the frame, seals the panel with both interior and exterior sheathing and installs windows and doors. Other panelized homebuilding options include structural insulated panels (SIPs) and pre-cast concrete. Panels are produced in the factory and transported to the home site where they are set onto a permanent foundation using a crane. The home is then finished by the homebuilder.

Like modular homebuilding, HUD Code (sometimes called "manufactured") homebuilding, uses large, three-dimensional, factory-produced elements to build a home. However, the HUD Code element (termed a unit or floor) must be constructed on a permanent chassis (usually steel) designed for over-the-road transportation. At the home site, the units are often placed on a non-permanent foundation, such as block piers. HUD Code homes are almost always single-story. The completed HUD Code home must comply with the federal HUD Code, which supersedes state and local building codes regarding design and construction. While HUD Code homes can be located in any state, state building codes and local zoning regulations severely limit where HUD Code homes can actually be placed.

Figure 1.3 Pictorial summary of modular homebuilding process

Although whole-house prefabrication has been proven to be operationally feasible for building conventional-size homes, it has yet to be proven financially viable. A major complicating factor has been the logistics of moving complete prefabricated homes from the factory to their home sites. A recent effort used a very large carrier that was restricted to operating over roadways in a single housing development adjacent to the whole-house factory. These transport restrictions limited the market served by the factory and contributed to the firm's failure.

1.3 THE PROMISE OF MODULAR HOMEBUILDING

Modular homebuilding holds great promise when compared to conventional site-built construction and other prefabricated homebuilding approaches. While maintaining a practical level of design flexibility, modular homebuilding allows more of the homebuilding process to be industrialized – relocated and transformed from many, temporary, open construction sites to a single, dedicated, enclosed homebuilding factory. All of the resources needed for homebuilding – the process, building materials, workforce and supporting infrastructure – are assembled and integrated in the factory.

The most visible advantage of the factory is that it provides shelter for the homebuilding process, which: 1) prevents process delays due to adverse weather; 2) shields the in-process home, building materials and equipment/tools from weather-related damage, vandalism and pilferage; and 3) protects workers from the elements. Another highly visible advantage of the factory is that it allows homebuilding activities to be performed in parallel, reducing overall production cycle time. For example, pre-set site work and foundation construction can take place at the same time that the factory is planning and producing the modules. Within the factory itself, production is massively parallel – wall assemblies are produced at the same time as the floor, the roof assembly is produced at the same time that the wall assemblies are being set, etc.

The factory provides much more than shelter for the homebuilding process. Conventional site-built construction processes are inherently variable, leading to process instability and erratic construction performance. This results from the nature of the workplace (multiple, temporary, open construction sites) and from the homebuilder's supply chain strategy of relinquishing the homebuilding process to independent subcontractors. The factory enables the development of a dedicated manufacturing process that is standardized and rationalized to provide safer, higher precision, timelier and more efficient homebuilding. A progressive assembly line is the backbone of the manufacturing process. It serves to integrate, coordinate and control the many fabrication, assembly and finish activities comprising the homebuilding process. This is achieved by creating a continuous flow of value-added activity as the module flows down the line. The line also synchronizes factory production with homebuyer demand, providing a highly visible pacing mechanism for each workstation, each work team and each worker in the factory. Workstations are set up along the line, each dedicated to a series of homebuilding activities. Specialized tools and equipment are provided in each workstation so that workers can perform their assigned tasks safely, precisely, quickly and efficiently. Examples include specialized framing jigs used to build floor, wall and roof assemblies and mechanized material handling equipment used to move large, bulky, heavy materials and subassemblies. Tools, equipment and building materials are located in the workstation for ease of access near their point-of-use. The method for each production activity is standardized and used as the baseline for continuous improvement. This tightly integrated production system consistently produces high quality homebuilding components safely, quickly and efficiently.

The management of building materials is also enhanced in the factory. Thousands of individual parts representing hundreds of unique building materials are used to build a home. These materials must be ordered, received, stored, staged for production, and used. After use, scrap and packaging materials must be collected and reused or disposed. For conventional site-built construction, these activities are scattered over many building sites, construction vehicles and staging yards. They are performed by a variety of personnel including site supervisors and subcontractors. In the factory, these activities are standardized, assigned to specific workers, and take place in designated areas. Storage systems are provided to enhance safety, material protection, accessibility, storage efficiency, and visibility. The visibility of commonly used items facilitates annual physical inventory and makes it easy to determine when and how much to reorder. Lean "pull" inventory replenishment systems can be instituted to further simplify inventory control. Consolidated "safety" stocks are reduced, while providing superior protection against delivery disruptions or quality problems. Each building material is received and inspected at a single location near its point of storage and use, enhancing incoming material quality and handling efficiency. When ready for use, materials are moved a short distance to their designated staging area near their point-of-use on the production line.

Waste management is enhanced in the factory. Less waste is produced since building elements are designed to take advantage of standard material sizes. The offal that is produced in fabrication is consolidated and used elsewhere in the factory. The remaining waste is more likely to be consolidated and recycled, increasing resource sustainability, reducing landfill needs, and reducing related costs.

The ready availability of a highly skilled, dedicated workforce is an important advantage of factory homebuilding. Tommelein [5] has described work on the construction site as a "parade of trades". Almost no homebuilders who build more than 50 homes per month actually perform any construction work [6]. Instead, they rely on 25–30 independent trade subcontractors who actually build the home. Hiring subcontractors is attractive to the site builder because their equipment and their workers are specialized to provide better quality and efficiency. It also provides financial flexibility, since the builder need not recruit, develop or maintain the workforce, or pay benefits. However, there are downsides to this subcontracted workforce [7]. Most subcontractors work for more than one builder. These builders have their own project demands and schedules. Therefore, subcontractors move frequently between different projects for different builders. Complicating this movement is the fact that each subcontractor depends upon the work of other subcontractors. Before a subcontractor can begin work, the work of preceding subcontractors must be completed to specification. Inspections by public building inspectors may also be required to certify that the work has been completed properly. This is often the case with structural framing, electrical, plumbing, and heating and cooling work. Effective scheduling, even under perfect conditions, requires close synchronization and coordination to assure a smooth transition as the many subcontractors "flow" through the construction site. The construction site, however, is not always a perfect workplace. Disruptions, delays and rework are common due to weather, material, workmanship, tools/equipment, design errors/change orders and myriad other issues. These often prevent a subcontractor from completing a job on schedule – and the rescheduling ordeal begins. The factory, on the other hand, is designed so that all homebuilding activities are synchronized. Delays and disruptions, though possible, are minimized. When they do occur, the factory responds as an integrated whole: temporarily moving cross-trained workers to a trouble spot, working overtime, and always asking why the problem occurred and what can be done to prevent its recurrence.

The capability of the factory worker is also an advantage. Site construction workers can be classified as general laborers or skilled tradesmen. In a recent career ranking, a career as a construction worker (laborer) was ranked 190 out of 200 [8]. Strieber [8] comments: "to wind up in the bottom 20 of our 200 overall rankings, a job must suffer from several glaring deficiencies such as high stress, challenging physical demands and low pay." He continues: "a majority of the 10 worst jobs in America require little more than basic training and, in most cases, a strong back and large muscles. Most pay less than $40,000 per year." Although various skilled trades were ranked higher in the ranking, all were ranked in the lowest one-third of careers. A recent survey revealed that only six percent of high school students hope to have a career in the skilled trades – defined as plumbers, carpenters, electricians, heating, ventilation or air conditioning installers, or repair people [9]. These results suggest that even the more desirable construction trades have difficulty recruiting and retaining talented workers. Because of these perceptions/realities, construction jobs have been increasingly relegated to the least capable – those who are unskilled, non-English speaking and sometimes undocumented. Many of these workers are part-time or even day laborers, without any job security or benefits.

This stands in stark contrast to a factory worker. Factory workers are treated as critical, long-term assets. They typically receive both classroom and on-the-job training. Training begins at hiring and continues periodically throughout the worker's career. Training includes both homebuilding trade skills and higher level production topics such as quality management and lean production systems. Cross-training produces a more well-rounded, flexible and valuable worker. Daily work assignments are highly focused. A factory worker typically has one to two hours to complete an assignment on a module, allowing a high level of worker specialization and resulting in higher labor productivity. Factory workers are well equipped with highly specialized tools and equipment, enabling safe, precise and efficient production. Convenient placement of material coupled with supportive material handling equipment minimize the need to man-handle large, bulky, heavy materials. For their efforts, factory workers receive competitive industrial compensation packages including wages, benefits (including health insurance), and often bonuses that are tied to company profitability.

A final advantage is that the factory provides an on-hand, supportive infrastructure for the production process. Three key elements of this infrastructure will be addressed: management, engineering and quality. The operating management structure of a site-built homebuilder typically consists of the builder and one or more site supervisors, each responsible for work on multiple construction sites. The supervisors spend a great deal of time scheduling and coordinating trade subcontractors that do not normally communicate with each other [7]. This limits management presence on each site. Once on site, the site supervisor's active management participation is indirect at best. The site supervisor must communicate through subcontractor foremen rather than directly to individual workers. This makes it more difficult to manage and focus the work.

The organization chart for a typical homebuilding factory shows several tiers of management. At the lowest level, a lead person is embedded in each work team. These experienced and highly skilled hourly workers not only perform routine production activities, but provide real-time guidance to the team – making team assignments, interpreting production drawings and coordinating on-the-job training for new workers. Foremen/supervisors are the first level of full-time, salaried management. Each is formally responsible for the performance of multiple work teams. A production manager is responsible for all production activities in the factory. A general manager is responsible for all operations in the factory including sales and marketing, engineering, production, finance and accounting, information systems and human resources. Members of the management team hold daily and/or weekly production meetings to review production performance, address issues, and plan for the future. It is not unusual for members of the management team to observe the process daily and to participate in final inspection of completed modules. The continued presence and active involvement of this well-organized management team are key contributors to the success of the homebuilding factory.

There is also synergy in co-locating production and engineering. Even the simplest standard home models have many design details. Most homebuyers demand customization, requiring additional and often more complex design details. Not all of these details are well documented in the production drawings. Workers on the construction site do not enjoy ready access to engineering staff, leading to delays during issue resolution or, more likely, ad hoc improvisation. The factory provides ready access between production and engineering, encouraging communication about issues, speeding their resolution, and preventing their recurrence. Ideally, production staff reviews new orders prior to release for production and requests clarification from engineering when needed. Production staff may also suggest that engineering revise details to simplify production. If a question arises during production, engineering is immediately consulted for clarification.

Quality is the responsibility of all workers in the modular factory. Each worker is also considered a customer, responsible not only for his/her own work, but for checking/inspecting previous work. This philosophy is formalized in some factories, with each work team signing-off on incoming quality before beginning its own assignment. Several organizations, both internal and external, are formally responsible for assuring quality in the modular factory. An internal quality control department is responsible for inspecting each module at selected locations on the line. Extensive checklists ensure compliance with applicable building codes and quality of workmanship. Quality control staff also perform a final inspection of each completed module to assure satisfaction of the homebuyer. Upper level factory management typically participates in these final inspections. Inspection results are communicated throughout the organization to initiate rework, institute corrective measures to prevent recurrence, and to track ongoing quality performance. Later inspection results from the homebuilder upon delivery and the homebuyer at closing are also reviewed and reported. In some factories, worker bonuses are affected by measured quality performance.

Independent, third party agencies play an important role in assuring quality in the modular factory [7]. Most states have a design approval and production inspection system for modular homebuilding. Quality systems typically include requirements for quality control procedures and manuals, a design review for each module produced and in-plant inspections by state-approved third party agencies and/or the state itself. A preemptive state code based on a model code (International Building Code) is most often used. In-plant inspection requirement vary. Some states require each module to be inspected at least once. Others require a small percentage of modules to be inspected, based on the assumption that quality assurance plans and repetitive manufacturing processes will prevent problems.

Site-built homebuilding is usually not subject to same degree of design review as modular homebuilding, nor do site builders have to adhere to formal programs that demonstrate that the construction process incorporates accepted quality improvement and control procedures [7]. Applying quality control on the construction site is complicated by the use of subcontractors and the constant change of skill levels of workers, the variety of home designs and options offered, and the often small number of homes built by a builder. As a result, site-built homebuilding quality relies on inspections by internal site supervisors and public building inspectors. Site supervisors oversee all aspects of construction on multiple construction sites and are also responsible for construction schedule and cost – an obvious conflict with their role of assuring quality. Although public building inspectors play a crucial role in assuring quality in site-built homebuilding, "in reality many inspections are very cursory and may not find all defects due to the wide variety of possible deficiencies and large number of inspections required of any given inspector." [7]

Modular homebuilding not only holds the promise for using fewer resources, but also the potential to reduce the basic unit costs of resources that are used. Factories located in areas where prevailing wage rates are lower achieve a competitive advantage by selling homes in markets where the construction wage rates are much higher. The large size and purchasing power of a few, multi-plant, modular firms enable them to negotiate substantial volume discounts for commodity-type building materials. These producers achieve savings through large-scale purchases direct from manufacturers instead of distributors or wholesalers, and by taking delivery at centralized production facilities rather than at multiple building sites. Savings are said to average 10% and may be as much as 30% [7].

1.4 THE REALITY OF MODULAR HOMEBUILDING

1.4.1 Encouraging signs

Modular homebuilding results have been encouraging for several key performance metrics. Bashford [10] estimated that the construction cycle time for 23 large publicly traded U.S. homebuilders, all conventional site builders, averaged 152 days in 2001. This estimate is consistent with the average cycle time actually observed for new site-built homes in Chandler, Arizona during the first six months of 2001. Bashford attributes this lengthy construction cycle to non-productive time occurring during the construction process. A multi-year study of site-built homebuilding operations in the Phoenix metropolitan area found that actual construction operations consumed 25% to 40% of available working time – leaving unfinished homes sitting idle over 50% of the available work time [11]. The comparable cycle time for modular homebuilding is typically 34 to 98 days: three to six days to produce modules in the factory while the foundation is being prepared on site, one to two days to set modules on site, and one to three months to finish the home on site. These findings suggest that, at least for the homebuilding activities that have been industrialized, much of the wasted idle time has been eliminated. These findings are important, both to the homebuilder who ties up working capital and pays carrying costs over an extended period of time and to the homebuyer who must wait for the new home and pay a higher price to cover the costs associated with builder delays. It should be noted that these estimates of construction cycle time for both site and modular homebuilders do not include pre-construction activities such as sales, financing and engineering and delays due to backlogs.

Researchers at the National Association of Home Builders (NAHB) Research Center estimated that the cost for a builder to construct the structure (including foundation) for a new 2,000 square-foot site-built home is about 18% more than that for an identical modular home [7]. After adding non-structure costs (such as land, site prep, overhead/general expenses, marketing, sales commission, profit, construction financing), the total sales price for the site-built home is estimated to be about 10% more than that of the modular home. These findings are important to the builder, who can use the sales price differential to increase market share, and to the homebuyer, who can reduce the cost of homeownership. The builder may also choose to take some of this differential as profit, sharing the cost benefits with the homebuyer. During periods of higher demand, modular producers may also seek a share of the differential, raising prices to "meet the competition" and translating their lower production cost into higher margins. This is particularly important since modular producers shoulder the burden of carrying higher fixed assets through downturns in the housing cycle.

It is instructive to explore the source of this cost differential, starting with the structure. The builder's cost to construct the structure includes material and labor. For a modular home, structure cost also includes the cost to purchase, ship, set and finish the modules. The purchase cost of the modules is inclusive of the modular producer's costs, which are shown in Table 1.1. The modular producer adds 85–95% of the value to the structure in the factory.

Table 1.1
Cost structure for typical modular
producer [12]

Cost Line Item	% of Sales Price
Materials	45–50
Overhead	35–45
Labor	10–20

Building material is a major cost line item in homebuilding. The modular producer provides most of the materials (excluding foundation, garage and deck) used to construct the structure for a modular home. Material costs represent 45% to 50% of the modular producer's sales revenue. Modular and site-built homes have similar Bills of Material, with modular homes requiring additional structural components to accommodate the structural stresses encountered during shipping and installation. As for pricing discounts, although a few, large, multi-plant, modular producers have been able to negotiate substantial volume discounts for commodity-type building materials, most modular producers have not grown large enough to realize this savings [7]. Therefore, it is reasonable to assume that lower material cost is not the primary driver of the lower cost for a builder to construct the structure for a modular home.

Labor is also a major cost line item in homebuilding. Labor costs have been estimated at 40% or more of total site-built home cost [7]. An industry-wide benchmarking effort led by the Systems Building Research Alliance found that labor costs (direct and indirect) for a modular producer consume 10% to 20% of sales revenue, averaging 16% [13]. Even considering the additional labor required to transport, set and finish a modular home, this strongly suggests that the lower labor cost in the modular factory is a primary driver of the lower cost to construct the structure for a modular home.

To attain a high level of labor productivity in the factory, a modular producer incurs substantial overhead costs: 1) factory capital costs for land, facilities and equipment and 2) ongoing support costs for the overall modular production enterprise. These overhead costs (35% to 45% of sales revenue for a typical modular producer) are passed along to the builder in the module purchase price and, thus, increase the cost to construct the structure for a modular home. These costs are not incurred by a builder constructing a site-built home.

To summarize, even with the offsetting costs for additional structural materials, unique post-manufacturing modular activities (transport, set, and finishing) and factory-related overhead, the builder's cost to construct the structure for a site-built home remains 18% more than that for an identical modular home.

Non-structure costs also contribute to the overall cost advantage of modular homebuilding. For example, the construction financing cost for a site-built home is estimated to be more than twice that of the modular home, reflecting the longer construction cycle time [7].

The physical safety measures incorporated in the modular factory combined with the development of a "safety first" workforce culture should result in reduced injury rates. According to data reported by the U.S. Bureau of Labor Statistics [14], the residential construction industry (site builders) experienced an injury rate of 4.7 injuries per 100 employees in 2007. The corresponding rate for the reporting category used by modular producers, prefabricated wood building manufacturing, was 2.3 injuries per 100 employees, less than half that of the site builders. It should be noted that even though modular producers are predominant in their category, the category also includes panelizers, log home manufacturers, and other prefabricated wood building manufacturers. Therefore, injury rates for modular producers may differ somewhat from the category average.

Industrialization should lead to improvements in materials management. Inventory turnover (the annualized cost of goods sold divided by average inventory value for raw materials, work-in-process and finished goods) is a commonly used metric for assessing materials management performance. As shown in Figure 1.4, modular producers average 13 inventory turns per year, better than that reported for other industries (six in construction, six in aerospace, twelve in automotive, and eight in durable consumer products) [13].

Figure 1.4
Inventory turns per year for modular plants [13]

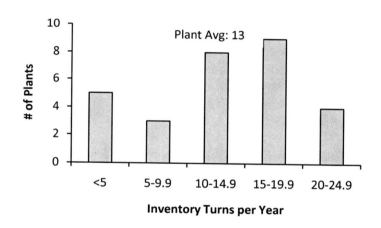

Laquatra and Pierce [15] performed an audit of construction waste during the on-site construction of an 1,894 square-foot single family home in upstate New York. The total weight of all waste materials was 2.3 tons. Gypsum, wood and cardboard waste made up almost 75% of total debris by weight. That figure was compared with seven other waste audits that have been conducted around the country and found to be consistent. The issues of reducing and recycling construction waste are important for both environmental sustainability and housing affordability. From the affordability perspective, builders pay twice for construction materials that could be recycled but end up in landfills: payment is made when the materials are purchased and fees are assessed when the materials are dumped [16]. These costs are then passed on to homebuyers in the form of increased home prices. Although no comparable data is available for modular homebuilding, extensive observations in numerous factories suggest far less waste. Most factories bundle cardboard for recycling. Wood waste is minimal, since larger scraps from most cutting operations are used elsewhere in the factory. Drywall installers are careful to install drywall so that scrap is minimized, since there is only limited opportunity for using all but the largest drywall scraps. There is no drywall recycling.

Industrialization should lead to enhanced quality and, thus, higher customer satisfaction. In a survey of home buyers in central Florida who bought new site-built homes in 2001, Nahmens [17] found that 64% were satisfied with their builders, 26% were not satisfied, and 10% were neither satisfied nor dissatisfied. Customer satisfaction for modular producers industry-wide averages 90% (Figure 1.5), much better than that of the central Florida site builders and some other industries (79% manufacturing/durable goods, 80% automotive) [13]. In a study commissioned by the modular industry, JD Powers surveyed homebuyers that bought modular homes from major modular producers in 2005 [18]. Those homebuyers who lived in their modular homes for six to nine months were less satisfied with their home than buyers of site-built homes. However, those living in their modular homes for more than 18 months were more satisfied than site-built homebuyers. JD Powers links these findings to the time to resolve problems in modular homes after move-in: 62% of reported problems required one visit, 23% took two visits, and 15% took three or more visits. The longer it takes for problem resolution, the longer the customer is not satisfied.

Figure 1.5 Profile of customer satisfaction ratings for modular plants [13]

1.4.2 Modular production statistics and market share

Recent U.S. housing production trends are shown in Figure 1.6. The data indicate that modular homebuilding has maintained its overall market share of two to three percent throughout the rise and fall of the housing market. Through the same time period HUD Code homebuilding has lost both production and market share. This continued deterioration in the HUD Code market has resulted in a major shift for some of the largest industrialized homebuilding companies. Three of the largest HUD Code manufacturers are now the three largest modular producers [22]. The largest modular producer produced 40% of its $1.4 billion revenue in 2006 from the modular business compared to just five percent in 2002 [21].

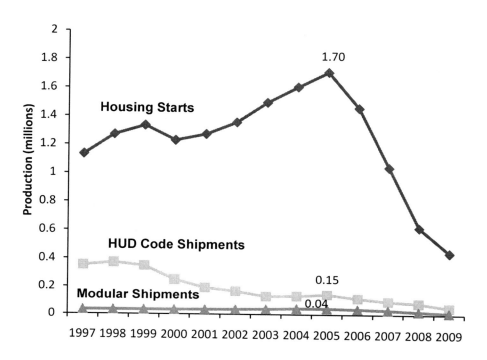

Figure 1.6 U.S. housing production trends [19,20,21]

A profile of new home construction in the U.S. by homebuilding technology is shown in Table 1.2. Although the data used to develop the profile are drawn from a variety of sources from 2002 through 2006, the results are believed to be reasonable estimates of market share. Fifty-five percent of homes were site-built using conventional wood frame construction (stick-built). Fourteen percent were site-built using concrete block. Thirteen percent were built using prefabricated panels – primarily low value added, open wood frame panels. Nine percent were HUD Code. Insulated concrete forms (ICFs) and removable concrete forms were used to build four percent of homes. Two to three percent were modular. Steel framing was used to build two percent of homes. Non-traditional, innovative technologies, such as structural insulated panels (SIPs), were used to build the remaining two percent of all homes.

1.4.3 Remaining challenges

Although the evidence suggests that modular homebuilding has provided some important benefits to both homebuilders and homebuyers, the sobering reality is that it has not lived up to its promise. This is evidenced best by market share, which has remained flat, averaging between two and three percent of overall single-family home production (Figure 1.6). Many explanations have been offered. Some have cited design issues. While modular home design has improved dramatically with innovations such as the folding roof truss (allowing a steeper roof pitch), it has not kept pace with site-built competitors in providing a sufficiently wide range of form, finish, detail, and technology options [1]. Some conjecture that when modular producers have customized their product to compete in a more upscale market, they have done so indiscriminately, lacking focus on the specific portion of the upscale market they wish to pursue [7]. This has raised the cost of modular homebuilding to near that of site-built, nullifying the potential advantages of industrialization. Some note that modular home designers/architects have failed to gain strategic advantage over site builders by identifying and incorporating key differentiating features that add great value for homebuyers and are particularly suited for factory production [27].

Table 1.2 Profile of homebuilding technologies [19,20,21,23,24,25,26]

Technology	Market Share
Stick-built	55%
Concrete masonry	14%
Panelized	13%
HUD Code	9%
Insulated Concrete Forms	4%
Modular	2–3%
Steel Frame	2%
Structural Insulated Panels	<1%
Other	<1%

From a design/materials standpoint, modular producers concede that modular home designs require more framing materials than comparable site-built designs. This assures that modules can withstand the stresses of shipping and crane handling during site installation. Examples include double rim joists used in floors, the marriage wall needed in each module and the extensive use of OSB sheathing to cover exterior walls. Several producers have indicated that a typical modular home uses 10–25% more material than comparable site-built homes [28], although this estimate appears overstated, even for framing and sheathing materials. Although some producers claim that the additional materials produce a stronger home, better able to withstand severe storms, the materials reduce the cost and sustainability advantages of modular homebuilding. Some also believe that modular home design has not fully capitalized on the engineering capabilities and standard dimensions of building materials [1], missing opportunities to limit labor and material waste.

Design is important to the homebuyer. The JD Powers survey of modular homebuyers [18] found that 77% of modular homebuyers indicated that the overall design and floor plan was extremely important in the selection of their home, the second highest factor behind quality of construction and workmanship. In a broader survey of all new home buyers, JD Powers [29] found that buyer satisfaction with new-home design is driven by the following factors (in order of importance): floor plan, master/primary bathroom, kitchen and the buyer's ability to customize the overall design. The survey also found that the flexibility of builders to make non-standard design changes – such as relocation of an interior wall – is particularly important in satisfaction with new-home design.

Marketing issues continue to plague the industry. Many homebuyers still associate modular homes with less expensive HUD Code homes and even with the older generation of "mobile homes." Modular marketers are torn between 1) recognizing this widely held perception and downplaying the factory-built aspect, positioning themselves as equivalent to site builders or 2) ignoring the risk of association and embracing the benefits of industrialization, differentiating themselves from site builders.

Regulatory requirements have been cited as an issue for modular producers. Although some states do not have special building requirements for modular homes, most modular homes are sold in states with unique regulatory requirements for modular homes. These requirements typically include the need for quality control procedures and manuals, a design review for each module produced and in-plant inspections by state-approved third party agencies and/or the state itself. Perhaps a larger regulatory issue for modular producers is the need to deal with multiple regulatory authorities and satisfy a variety of building codes to sell in a large geographic area. Since most modular homes are located in states that have adopted a preemptive state modular building code, statewide access has become less of an issue. Yet even after years of working for effective reciprocity agreements, the process of design approval and in-plant inspection of modular homes destined for interstate shipment remain duplicative and inefficient, due to limited coordination across state lines [7]. This has slowed the growth of modular producers and limited their ability to attain economies of scale.

Modular producers have failed to take advantage of modern manufacturing technologies that can vastly improve quality, cycle time and productivity. Much of this failure is linked to the choice of wood frame construction as the predominant building system. The long-term performance of wood frame construction is well known and it is widely accepted for homebuilding by the general public, regulatory authorities and model building codes. However, it dictates a set of building materials and components that were designed for low technology construction processes on the building site, not for modern manufacturing processes. Elemental processes used by modular producers are essentially manual and remarkably similar to their site-built counterparts, who largely use the same materials. Low technology mechanized hand tools such as pneumatic nailers and paint sprayers are common. Critics have observed that modular producers still "stick build under a roof", referring to the industry's continued reliance on these relatively unsophisticated building materials and their associated low technology processes. While the choice of building system undoubtedly limits the performance of the modular factory and the homes produced there, one should not underestimate the potential for the factory to add additional value and eliminate waste. As described earlier, the factory does far more than provide shelter for conventional site-built homebuilding processes.

Recognizing the need to gain a competitive advantage over site builders who use the same building system, modular producers have sought new materials, equipment and systems engineered specifically for factory production. Because of their relatively small size, suppliers have not focused on the modular market. For example, automated wood frame panelizing equipment has been designed for panel producers who operate at much higher rates. Modular producers, with their lower volumes, have not been able to justify the more expensive, higher capacity equipment [27].

When suppliers do develop promising new materials, equipment and systems that offer greater precision, capacity and efficiency, modular producers are often unable or reluctant to invest. Modular producers are usually smaller and less well capitalized than other manufacturers [7]. Even when capital can be obtained, innovation and expansion are tempered by the risks. The greatest risk for highly leveraged industrialized homebuilders is a major downturn in the highly cyclic housing market, which is largely driven by mortgage interest rates and, more recently, by the availability of mortgage credit. Modular producers are wary of making heavy long-term financial commitments to capital facilities, equipment and systems, particularly when sales can be devastated by unexpected events [30]. A secondary risk is the long-term commitment of suppliers that do develop new products specifically for modular production. For example, several plants were recently outfitted to handle a new supersize drywall product (8' wide and up to 24' long). The price of the new product rose substantially and the expected demand never materialized. The product was eventually pulled from the market, leaving innovative manufacturers with sizable investments that could not be recovered [27].

A multi-year, industry-wide joint effort led by the Systems Building Research Alliance (SBRA) found that there is considerable opportunity to improve the efficiency of existing factory-based homebuilding processes. One of the first tasks of the SBRA effort was to assess the current state of the industry. An industry-wide benchmarking survey was performed covering a broad range of performance metrics [13]. Among other findings, the survey revealed sizeable performance variation between modular plants. A modular plant is defined to be a factory in which modular homes represent more than 50% of production. A profile of direct labor productivity is shown in Figure 1.7. Even after considering differences in regional labor cost, product mix, and customization, the size of the variation in labor cost suggests considerable opportunity for many modular producers to improve labor efficiency to best-in-class performance levels.

Figure 1.7
Profile of direct labor productivity for modular plants [13]

A second phase of the SBRA effort demonstrated that the introduction of lean production techniques can provide striking efficiency, quality and other improvements for existing housing plants [31]. Participating production departments in nine plants experienced productivity improvements ranging from 10% to more than 100%. Quality improvements included one plant where defects in the finished drywall process were reduced by 85%. Clearly, much more can be done in factory organization and operation to increase the benefits associated with modular homebuilding.

A surprising finding revealed by the SBRA benchmarking study was that many modular plants were operating at production rates that were well below capacity (Figure 1.8). Plant capacity averaged 66%, compared to 70% to 75% in other industries [13]. This is particularly problematic for modular producers, since they carry heavy overhead fixed costs that are typically 35% to 45% of sales revenue. Any reduction in production volume increases the cost per home, reduces the profit per home and reduces the cost competitiveness of modular homebuilding. There are several possible explanations for the lowered production rate. The first is that most modular producers were not able to sell enough homes to reasonably load their factories, either because of overall market demand or because of the design and marketing issues described above. Since the survey was performed in 2004 during a relatively strong phase of the housing cycle, it is unlikely that overall demand limited modular sales. Although design and marketing issues may have limited sales, there is likely a secondary cause.

Figure 1.8 Capacity utilization for modular plants [13]

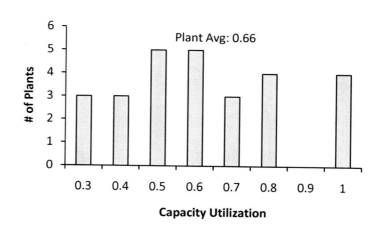

It is likely that some modular producers have intentionally (or unintentionally) limited their production rates to better accommodate increasing product mix and customization. As homebuyers become more sophisticated, modular producers are routinely customizing standard models and increasingly building one-of-a-kind custom homes, multi-family condominiums and apartments. Modular manufacturers may actually be offering greater product choice than some production builders, who have been intentionally narrowing design options. In an effort to quantify the impact of product choice on factory operations, Nahmens [32,33] found that producers offering increased product choice are likely to suffer poorer labor productivity, greater inventory, higher production costs, lower space efficiency, more quality issues and less satisfied homebuyers. In general, operational performance deteriorated with an increase in product choice. Therefore, industrialized housing producers have not reached the ideal of mass customization – designing and manufacturing customized products at mass production efficiency and speed. Instead, modular producers and their customers are paying a price for increased product choice.

The most immediate impact of increased product choice (particularly customization) is an increase in the engineering workload, which can eventually create a bottleneck. A bottleneck is the slowest operation in a process, the operation that limits production capacity [34]. The engineering bottleneck can be readily, but not cheaply, managed by increasing capacity – increasing efficiency, working overtime, adding staff or outsourcing. However, identifying and managing production bottlenecks are more difficult. In the SBRA benchmarking study, modular producers noted the activities/plant features that bottlenecked their production (Figure 1.9). Roof and wall framing were cited most often as bottlenecks, followed by the line configuration (stations and system), and drywall finishing. Most producers identified more than one bottleneck in their plant. Mullens [27] has conjectured that multiple "floating bottlenecks" are a primary culprit in reducing plant capacity. A floating bottleneck is a bottleneck that shifts between activities, depending on product mix [35].

Figure 1.9
Bottleneck
activities/features
noted by modular
plants [13]

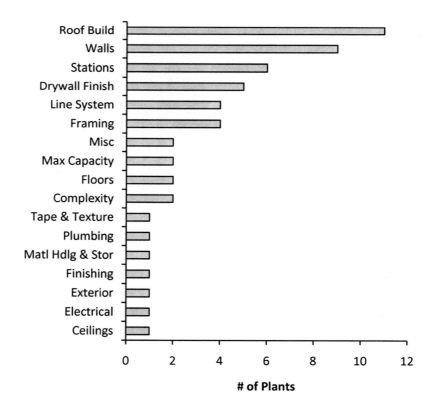

Bottlenecks interrupt flow, a primary advantage of industrialization. If the bottleneck is a feeder workstation producing major subassemblies for the line, such as wall or roof assembly, upstream modules cannot flow down the line. Since there are no queues, this quickly "chokes" the line and its feeder workstations. New modules cannot flow onto the line, and upstream work activities are delayed. Downstream of the bottleneck station, the line moves, creating holes on the line and delaying activities. When the bottleneck shifts to an activity that can move on the line, such as drywall finishing, the line can continue to cycle forward and upstream activities are not affected. However, downstream activities are still delayed. As delayed modules move further down the line, they move further away from staged materials and supporting workshops. This introduces additional inefficiencies. Eventually, unfinished modules can exit the still-cycling line and end up in the yard. This introduces huge inefficiencies and probable damage and rework. To prevent this, some producers will not allow incomplete modules to exit the line, thus creating another bottleneck at the end of the line. Limited work measurement studies in rough framing and drywall have indicated that delays represent at least 10% to 15% of work time and perhaps substantially more [27]. This does not include working time at reduced productivity levels. To compensate for the delays and to maintain schedule, producers work overtime at a 50% labor cost premium.

Delays are not the only consequence of floating bottlenecks. Instead of working at a steady, sustainable pace, workers routinely "hurry up and wait." Frustration and exhaustion are the natural consequences, and quality is a likely victim. Studies in modular plants have shown that at least seven percent of all labor hours are spent on rework, and 47% of all drywall labor hours are rework [36]. Rework cannot be expected to remedy all quality problems. Thus, at least some of quality problems found by the homebuyer (and the resulting service cost) can be attributed to floating bottlenecks.

In general, manufacturers that are challenged with a limited production rate and heavy fixed costs often add a second shift. This amortizes fixed cost over more production with only a minimal variable cost increase. Modular producers have chosen not to add a second shift. They attribute their decision to: 1) the difficulty in recruiting workers for the second shift, 2) a modest shift differential for labor rates and 3) the difficulty of managing the transition between shifts – where workers must successfully complete the work started by others on the previous shift. In some cases, producers have required their workers to work overtime over extended periods. However, there is a 50% premium in the overtime labor rate, and extended overtime has often generated serious worker dissatisfaction and contributed to labor turnover.

Although the dedicated workforce is an important advantage of factory homebuilding, modular producers have reported workforce issues. Senior executives report that recruiting and retaining their workers are two of their most critical problems. This issue is illustrated in the SBRA benchmarking results shown in Figures 1.10 and 1.11. Average annual labor turnover is 50% with some firms reporting more than 100%. These rates are greater than those in other industries (28% in construction, 17% in manufacturing) [13]. Average daily absenteeism is more than 5% with some plants reporting more than 10%. Again, these rates are greater than the industrialized sector of the economy overall (3%), but less than some industry sectors such as automotive (10%) [13].

Figure 1.10
Annual labor turnover for modular plants [13]

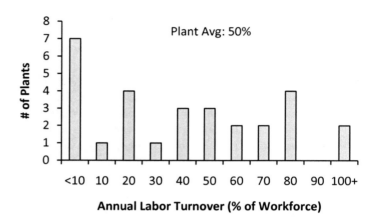

Figure 1.11
Average daily absenteeism for modular plants [13]

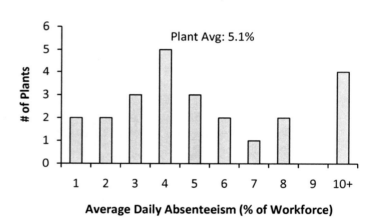

JD Powers [18] reported that 86% of modular homebuyers in their survey cited the quality of construction and workmanship as being extremely important in the selection of their home, the highest rated factor. This should be welcome news to modular producers, whose extensive quality efforts should yield a superior quality home. However, the findings are mixed. JD Powers [18] also found that those modular homebuyers who lived in their home for 6–9 months were less satisfied with their home than buyers of site-built homes. However, those modular homebuyers living in their homes for more than 18 months were more satisfied than site-built homebuyers. JD Powers links these findings to the time to resolve problems in modular homes after move-in: 62% of reported problems required one visit, 23% took two visits, and 15% took three or more visits. The longer it takes for problem resolution, the longer the customer is not satisfied.

No industrialized homebuilder has won the Malcolm Baldrige National Quality Award, the most prestigious national award for quality. Only one industrialized homebuilder, Palm Harbor Homes – Florida Division, has won the National Housing Quality – Gold Level Award. At the time of the award, Palm Harbor was primarily producing HUD Code homes. Elshennawy, Mullens and Nahmens [37,38] report that the modular industry has been reluctant to embrace the broad elements of quality management and go on to describe how a continuous improvement-based quality management system might be implemented. They found that modular producers say that quality is valued in their organizations and that producers do inspect their product aggressively on the production line. They also found that modular producers adhere to formal state mandated programs (where they exist) that are verified by independent third party inspectors and that must demonstrate that production processes incorporate accepted quality improvement and control procedures. However, they found that there is no coordinated pursuit of quality and that modular producers are certainly not obsessed with quality. Measured quality performance also indicates opportunity for improvement. Figure 1.12 shows the percentage of overall labor hours consumed by rework in a typical modular factory. The average over all activities is 7%, with a peak of 47% for drywall finishing. Service costs (the costs to repair problems found by the homebuyer) are also an important indicator of quality. SBRA benchmarking results for service costs are shown in Figure 1.13. While service costs average 2.8%, a number of plants spend more than twice that amount.

Homebuilding activities on the construction site represent between 5% and 15% of the modular homebuilding process. Yet it requires an inordinate amount of the overall homebuilding cycle time and is responsible for many quality problems. Mullens and Kelley [4] found that modular homebuilders still finish homes on the construction site using conventional construction paradigms, failing to capitalize on the opportunity to slash delivery times with a more lean approach. JD Powers [18] found that only 35% of modular homebuyers reported the home was ready for move-in by the date originally promised by the builder. In a broader study of all new home buyers, JD Powers [29] found that the proportion of homes delivered both completely finished and on time is 70% in 2008, more than twice that of modular homebuilding.

Progress made by competitors in the site-built industry, particularly by production builders, has eclipsed many of the gains made by modular homebuilders. The 23 publicly traded production homebuilders built over 20% of all homes in 2001 [10]. Production builders are large scale site builders, many publicly held, that seek to attain the efficiencies of industrialization while building on-site. Their volume has given them some of the same advantages enjoyed by factory-based industrialized homebuilders: construction sites that are located in close proximity, dedicated subcontractors, sophisticated quality management systems, and volume material pricing. Top executives at Pulte, one of the largest and most successful production builders, shared their homebuilding process simplification strategy at a recent investor conference [39]. In the short term it included reducing the number of floor plans, aggregating purchases, and reducing its overall supplier base. Long-term, Pulte remains committed to lean manufacturing and simplification of its supply-chain, which they believe will feature direct shipments of materials from centralized warehouses to facilitate a real-time and streamlined home construction process. The NAHB Research

Center has developed recommendations to accelerate the growth of production builders [7]. Several of these recommendations include collaborative relationships with factory-based industrialized homebuilders.

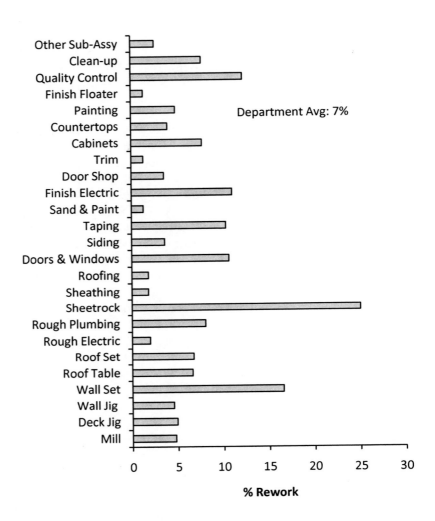

Figure 1.12
Rework as % of overall labor for a typical modular plant [34]

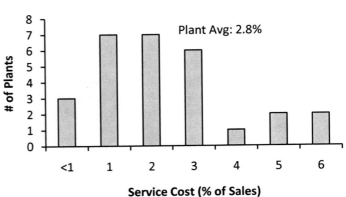

Figure 1.13
Service cost as % of sales for modular plants [13]

1.5 ATTAINING THE PROMISE

Modular producers have made significant strides toward attaining the promise of modular homebuilding. However, these strides have been too little to impact the basic momentum of the housing marketplace. Market share has been stagnant. The challenges that remain are daunting. Some challenges are external – difficult regulatory environment, inattentive suppliers, escalating market cycles and growing capability of production builders. Other challenges are internal – how the factory is configured and operated by a producer. Still other challenges are mixed – the capability of modular homebuilders (homebuilders who build with modules) and their relationship to modular producers. The overall objective of this book is to assist every modular home producer as it endeavors to attain the elusive promises of modular homebuilding by capitalizing on its strengths and confronting the difficult challenges. The focus is the configuration of the modular factory, not the bricks and mortar, but the production system. The principles apply to the design of a new greenfield factory, the selection and redesign of an existing facility, or the redesign of an existing modular factory.

The remainder of the book is organized as follows:
- Chapter 2 – uses the key findings from Chapter 1 to develop a production strategy that guides factory design. Lean production and mass customization are prominent in the production strategy.
- Chapter 3 – describes the primary building elements used in modular homebuilding and the production processes used to produce them.
- Chapter 4 – introduces a six-step structured engineering design approach for configuring the modular factory.
- Chapter 5 – discusses the implementation of the modular factory.

It is important to have a motivating vision before one begins a difficult quest. The following admittedly idyllic vision is humbly proposed as a starting point [27].

The factory will produce high quality custom homes at a competitive price for all homebuyers, from entry level through luxury. The factory will provide a productive and safe environment that will offer excellent value and timely delivery for the homebuyer, a safe and rewarding career for employees and a profitable investment for owners. Ample capacity will be provided to accommodate forecasted short-term growth. Factory design will be modular and flexible to facilitate expansion to accommodate more rapid or longer-term growth. Materials will arrive in the factory just in time to support production and be staged close to the point-of-use on the line. Mechanization/automation will be provided for both material handling and manufacturing processes when justified to eliminate injuries, minimize excessive physical exertion, assure capacity and boost productivity. Production documentation will be timely and accurate. Employees will know the status of any order, recognize the restrictions at any workstation or work group, and be able to react so that schedule and customer demands can be profitably met. As a result, rework will be minimal and production flow will be smooth and synchronous with demand. Employee work groups will be actively engaged in continuous improvement and will share in the resulting profits.

A final caveat is in order before proceeding. A well-planned factory is not necessary nor is it sufficient for business success. Success is also highly dependent on the general health of the industry, the local market (homebuyer demand and homebuilding competitors), etc. More than one poorly equipped producer enjoyed financial success in the overheated market of the early twenty-first century, and more than one well-equipped producer has closed in the aftermath. However, a well-equipped producer is more likely to be highly profitable during good times and less likely to be burdened during bad.

1.6 REFERENCES

1. Carlson, D., *Automated Builder: Dictionary/Encyclopedia of Industrialized Housing*, Automated Builder Magazine, Carpinteria, CA, 1995.
2. Gianino, A., *The Modular Home*, Storey Publishing, North Adams, MA, 2005.
3. Mullens, M. "Results from Studies of the Module Set Process," presentation given to the Quality Modular Building Task Force, Annapolis, October, 2000. http://www.housingconstructabilitylab.com/pages/set_pres.pdf, May, 2009.
4. Mullens, M. and M. Kelley, "Lean Homebuilding Using Modular Technology," *Housing and Society*, 31(1), 41-54, 2004.
5. Tommelein, I., Riley, D., and Howell G. Parade Game: Work Flow Variability vs Succeeding Trade Performance, *Journal of Construction Engineering and Management*, ASCE, New York, 125(1), 1999.
6. Bashford, H., Sawhney, A., Walsh, K., and Kot, K. "Implications of Even-Flow Production Methodology for the US Housing Industry" *Journal of Construction Engineering and Management*, 129(3), 330–337, 2002.
7. NAHB Research Center, Inc., *Factory and Site-Built Housing: A Comparison for the 21st Century,* U.S. Department of Housing and Urban Development, Office of Policy Development and Research, Washington, D.C., October, 1998.
8. Strieber, Andrew, *JobsRated.com: A Comprehensive Ranking of 200 Different Jobs*, http://www.careercast.com/jobs/content/JobsRated_Top200Jobs, May, 2009.
9. Durk, K. *Survey of High School Students Reveals Careers in the Skilled Trades Are at Bottom of the List*, http://news.morningstar.com/newsnet/ViewNews.aspx?article=/BW/20090204005 137_univ.xml, May, 2009.
10. Bashford, H. (2004). "On-Site Housing Factory: Quantification of It's Characteristics" *Proceedings of the NSF Housing Research Agenda Workshop*, Feb. 12-14, 2004, Orlando, FL. Eds. Syal, M., Mullens, M., and Hastak, M. Vol 2, pp. Focus Group 1.
11. AzPath Report No. 4, "Process Mapping of Residential Slab Construction Processes, 2001, http:/construction.asu.edu/AZPATH/Azpath%20Report-Information.htm., September, 2003.
12. Mullens, M. "Shop Floor Information Systems for the Modular Factory," presentation given to the Building Systems Council 2003 Showcase, Homestead, VA, November, 2003.

13. Manufactured Housing Research Alliance, *Develop Innovations in Manufacturing Processes through Lean Production Methods*, U.S. Department of Housing and Urban Development, Affordable Housing Research and Technology Division, Washington, D.C., October, 2005.
14. *Incidence Rate and Number of Nonfatal Occupational Injuries by Industry, Private Industry, 2007*, Bureau of Labor Statistics, U.S. Department of Labor, http://www.bls.gov/iif/oshwc/osh/os/ostb1909.pdf, May, 2009.
15. Laquatra, J. and M. Pierce, "Managing Waste at the Residential Construction Site," *The Journal of Solid Waste Technology and Management*, 30(2), May, 2004.
16. Yost, P. and Lund, E., *Residential Construction Waste Management: A Builder's Field Guide.* Upper Marlboro, MD: National Association of Home Builders Research Center, 1996.
17. Nahmens, I. *Dimensions of Service Quality in Homebuilding.* Master's thesis, University of Central Florida, Orlando, FL, November, 2003.
18. Long, T., "J.D. Power & Associates Surveys Modular Customer Satisfaction," *Builder Systems Magazine*, March/April, 2007.
19. *New Privately Owned Housing Units Started: Annual Data*, Manufacturing, Mining and Construction Statistics, U.S. Census Bureau, *http://www.census.gov/const/www/newresconstindex.html*, September, 2007.
20. *Shipments of New Manufactured Homes: 1959–2006*, Manufacturing, Mining and Construction Statistics, U.S. Census Bureau, *http://www.census.gov/const/www/mhsindex.html*, September, 2007.
21. *NMHC's Quarterly Modular Housing Report: First Quarter Issue - 2010 Volume 1*, National Modular Housing Council, Arlington, VA, June, 2010.
22. "The 2006 Top 31 Modular/Whole-house Panel Builders", *Builder Online* webpage, http://www.builderonline.com/content/builder100/2006B100/2006B100ModularTop31.pdf, September, 2007.
23. "Concrete Posts Strong Market Gains", *Concrete Homes Newsletter*, November/December 2006, http://www.cement.org/homes/ch_newsletter2006-11&12.asp#Market, September, 2007.
24. *2002 Residential Market Data Report*, Steel Framing Alliance, http://www.steelframing.org/sfa_datastatistics.shtml, September, 2007.
25. Schwind, C., *News Release - 6/6/2006 #2*, Structural Insulated Panel Association, June 5, 2006. http://www.sips.org/content/news/index.cfm?pageId=175, September, 2007.
26. Merry, J. and Adair, C., "Forecast 2004: Structural Wood Panels and Engineered Wood", *APA Media Center News Release #: C4-2004*, March 18, 2004, http://www.apawood.org/level_d.cfm?story=1169, September, 2007.
27. Mullens, M., "Production Flow and Shop Floor Control: Structuring the Modular Factory for Custom Homebuilding" *Proceedings of the NSF Housing Research Agenda Workshop*, Feb. 12-14, 2004, Orlando, FL. Eds. Syal, M., Mullens, M., and Hastak, M. Vol 2, pp. Focus Group 1.
28. Cameron, P. and Di Carlo, N. *Piecing Together Modular: Understanding the Benefits and limitation of Modular Construction Methods for Multifamily Development*, Masters Thesis, Department of Architecture and Department of Urban Studies and Planning, MIT, Cambridge, MA, September, 2007.

29. J.D. Powers and Associates, *J.D. Power and Associates Reports: Overall Customer Satisfaction with Home Builders Improves Significantly in 2008, Despite Daunting Economic Challenges and Surplus Inventory Levels in the Industry*, Press Release, http://www.jdpower.com/corporate/news/releases/pressrelease.aspx?ID=2008145, May, 2009.

30. Mullens, M., " Innovation in the U.S. Industrialized Housing Industry: A Tale of Two Strategies," *International Journal for Housing Science and Its Applications*. 32(3), 163-178, May, 2008.

31. Manufactured Housing Research Alliance, *Pilot Study: Applying Lean to Factory Homebuilding*, U.S. Department of Housing and Urban Development, Office of Policy Development and Research, Washington, D.C., July, 2007.

32. Nahmens, I. and M. Mullens, "The Impact of Product Choice on Lean Homebuilding," *Construction Innovation: Information, Process and Management*, 9(1), 84-100, March, 2009.

33. Nahmens, I. *Mass Customization Strategies and Their Relationship to Lean Production in the Homebuilding Industry. Ph.D. Dissertation*, University of Central Florida, Orlando, FL, August, 2007.

34. Goldratt, E., *The Goal.* North River Press, Great Barrington, MA, 1992.

35. Hopp, W. and Spearmann, M., *Factory Physics: Foundations of Manufacturing Management*, 2nd Edition, Irwin/McGraw Hill, New York, 2001.

36. Mullens, M., *Milestone Report: Results of Industrial Engineering Studies of the Epoch Corporation Modular Manufacturing Facility*, Report to National Renewable Energy Laboratory, Deliverable 2.C.1 Task Order KAR-5-18413-02, December, 1998.

37. Elshennawy, A., M. Mullens, I. Nahmens, "Quality Improvement in the Modular Housing Industry" *Industrial Engineering Research '02 Conference Proceedings*, Orlando, May, 2002.

38. Mullens, M., "Quality Systems and Improved Plant Design," presentation given to the Quality Modular Building Task Force, Salem, MA, October 2001, http://www.housingconstructabilitylab.com/pages/qual%20pres%20[Compatibility%20Mode].pdf, May, 2009.

39. Dugas, R. and Cregg, R., "Notes From The Road: Staying Leveraged For An Eventual Upturn", Presented at JPMorgan's 3rd Annual Basics and Industrials Conference, June 5, 2008, https://mm.jpmorgan.com/stp/t/c.do?i=C23A-422&u=a_p*d_206047.pdf*h_-238ia12, May, 2009.

CHAPTER 2
A PRODUCTION STRATEGY FOR MODULAR HOMEBUILDING

Chapter 2 proposes a production strategy to guide factory design and operation. The strategy seeks to attain the promise of modular homebuilding by building upon current strengths and confronting the difficult challenges.

2.1 MODULAR PRODUCTION STRATEGY: OVERVIEW

A production strategy should continue to deliver and even enhance the benefits that are being achieved by modular producers, while aggressively confronting the difficult unresolved challenges. Perhaps the greatest of these challenges is how to further reduce the cost of modular homebuilding – enough to compel homebuilders and homebuyers to seriously consider modular homes. This cost reduction must be accomplished without sacrificing the quality of construction and workmanship, home design and ability to customize that are highly valued by the homebuyer. Given this goal, it is useful to revisit the cost structure of modular housing producers (Table 1.1). Building materials are the largest cost line item, consuming 45–50% of sales revenue. Therefore, a production strategy must include a focus on the supply chain. Working upstream in the supply chain, partnerships with material manufacturers and distributors should be developed to identify opportunities to reduce ordering and shipping costs, improve incoming material quality, and secure quantity discounts. Working downstream in the supply chain, partnerships with modular homebuilders should be developed to identify opportunities to shorten order-taking cycle time, increase order acceptance rate, reduce change orders, meet promised home-completion dates, improve finished home quality, and speed service time. In a study involving a factory-built home producer and selected members of its dealer network, researchers from the Systems Building Research Alliance found great potential to improve the home sales process using lean production techniques [1]. The same study also found promise in applying lean techniques to installation on the construction site. Mullens and Kelley [2] report long-term findings from a large modular homebuilder that applied lean techniques to finish operations on the construction site. The builder reported a 59% improvement in productivity and a 22% reduction in completion time.

Overhead (including profit) consumes 35–45% of sales revenue. Overhead costs account for startup capital as well as ongoing enterprise support costs. To control capital costs, value engineering should be used to justify factory design decisions involving land, factory facilities and production equipment. Surprisingly, capital costs may only contribute modestly to overhead. Ongoing enterprise support costs such as administrative salaries, utilities, and insurance may far exceed capital costs and should also be controlled. Since overhead costs are largely fixed, the overhead cost per home falls as home production rises. This provides a powerful mechanism to reduce costs – increasing production to spread overhead costs over more homes and thus reducing overhead cost per home. Therefore, the production strategy must include a focus on capacity.

Labor costs (direct and indirect) for a modular producer consume 10–20% of sales revenue, averaging 16% [3]. Comparable labor costs for site-built homes have been estimated at 40% or more of total cost [4]. This cost differential is a primary driver of the overall cost advantage for modular homebuilding and must be protected and enhanced. Ongoing efforts in numerous housing plants have demonstrated that lean production techniques can provide striking improvements in efficiency, speed and quality [5]. Participating production departments experienced productivity improvements ranging from 10% to more than 100%. Therefore, the production strategy must include a focus on lean production.

As modular producers become more aggressive in reducing costs, their rush to efficiency must not jeopardize the quality of construction and workmanship, home design and the ability to customize that is highly valued by the homebuyer. With regard to quality, factory rework is a source of considerable waste, even at current efficiencies. When defects are missed in the factory and discovered by the homebuyer, they result in far greater costs and, more importantly, homebuyer dissatisfaction. The introduction of lean production techniques in one modular factory resulted in an 85% reduction of finished drywall defects identified by factory inspectors [5]. This reinforces the need for lean production in the production strategy. However, the importance of quality is so high that a more coordinated and highly visible focus on quality is warranted. Therefore, the production strategy must include an explicit focus on quality. Mullens [6,7] reports on the status of quality management at modular producers and suggests approaches for implementing quality management systems.

Many modular producers offer considerable choice to the homebuyer. These producers offer a wide portfolio of standard models, and allow customers to customize their homes within broad limits. Many producers also build one-of-a-kind custom homes, multi-family condominiums and apartments. Research findings suggest that producers offer this choice at a significant loss in operational performance: lower efficiency, reduced quality and lower capacity. Therefore, the production strategy must embrace the concept of mass customization – designing and manufacturing customized products at mass production efficiency and speed.

In summary, the modular production strategy must encompass five major focus areas: supply chain management, capacity management, lean production, quality management and mass customization. Three of these focus areas – capacity management, lean production, and mass customization – directly impact factory design and are discussed in greater detail in the following sections.

2.2 PRODUCTION STRATEGY: KEY FOCUS AREAS FOR FACTORY DESIGN

2.2.1 Capacity management

Managing capacity can reduce costs for a producer. When demand permits, increasing production spreads overhead costs over more products and reduces overhead cost per unit. Factory capacity can be increased in several ways. Perhaps the simplest approach is by adding working hours to the factory schedule, either incrementally with mandatory overtime or by operating a second shift. Neither requires an increase in the basic production rate with its associated process changes and increased overhead costs. Mandatory overtime incurs a substantial overtime labor rate differential, but does not require an increase in the labor force. Depending on the extent/duration of the overtime and the nature of the labor force, mandatory overtime may improve or worsen worker satisfaction and employee turnover. In general, mandatory overtime should be considered only as a short term response to varying demand, rather than a long term solution to a sustained increase in sales. Adding a second shift can double plant capacity. The addition of a third shift can add still more capacity. There are disadvantages of operating more than one shift per day. Workers on the second and third shifts must be paid a modest shift differential, increasing labor cost per unit. It can be more difficult to hire and maintain qualified workers for the second or third shift. Workers on the second and third shifts may not have the same level of staff support (for example, engineering to resolve design issues). It may also be difficult to manage the transition between shifts – where workers must successfully complete the work started by others on the previous shift. As a result, productivity and quality may suffer. For these reasons, modular producers have resisted operating more than one shift. Despite these challenges, factories in many industries, particularly those that are highly capitalized, routinely operate a second shift.

Producers can also increase capacity by increasing the production rate or, conversely, by reducing cycle time – the duration of time required to complete a unit of work on a product. Cycle time is reduced by identifying those activities that, on the average, cannot be completed in the desired cycle time (bottleneck activities) and restructuring the activities so that they can be completed. This is accomplished by rationalizing the process to increase efficiency and then adding necessary resources such as labor, equipment and floor space. The goal is to decrease the overhead cost per unit (and, therefore, the total cost per unit) with the increase in production. As cycle time is reduced, the setup (transition) time between cycles becomes a larger percentage of the cycle. Therefore, a producer can expect diminishing returns as processes are rationalized, unless setup times are concurrently reduced. The need for increased production speed, decreased setup times and their enabling efficiencies further reinforce the need for lean production in the production strategy.

Phasing capacity growth over time should also be considered to synchronize capacity with growing demand and to minimize initial capital requirements. This may mean starting with a smaller factory and then adding capacity, as needed, by enhancing an existing line, adding a new line, or even building a new plant. This can be accomplished by providing expansion walls in the original factory and sufficient buildable land on the site to accommodate planned growth. Other plant infrastructure such as utilities and heavy equipment layout may also be flexibly configured to facilitate planned growth. It is important to reemphasize that the capital costs associated with the factory are only a modest part of total overhead. Therefore, it would be counterproductive to over-value engineer the factory to reduce capital costs, if it limited critical capacity expansion opportunities.

All of these approaches, individually and in combination, should be considered when contemplating a change in capacity level. To assess the impact of various capacity planning options, the total cost per unit should be a key metric. Other important measures should include the potential impact on quality, design flexibility and worker satisfaction.

2.2.2 Lean production

Originating with the Toyota Production System [8], lean production is the result of decades of development by automobile manufacturers, who have reduced the average labor hours per vehicle by more than half, with one-third the defects [9]. Other industries have followed the automobile industry's lead, achieving similar results [10,11]. The industrialized housing industry is a late adopter of lean production. Recent efforts have demonstrated that the introduction of lean production techniques can provide striking efficiency, quality and other improvements for existing housing plants [5]. Participating production departments in nine plants experienced productivity improvements ranging from 10% to more than 100%. One modular plant reduced defects in the finished drywall process by 85%.

The goal of lean production is to satisfy the customer by delivering the highest quality at the lowest cost in the shortest time. This is accomplished by continuously eliminating waste. All forms of waste (see Table 2.1) are targeted by lean production initiatives. Lean refers to both a general way of thinking and specific practices that emphasize using less of everything [12]. Toyota uses a house to summarize lean production concepts as shown in Figure 2.1.

Table 2.1 The seven types of waste

Waste	Description
1. Defects/Correction	Rework/replacement and related efforts to correct a mistake.
2. Overproduction	Producing more than what is currently needed.
3. Transportation	Unnecessary movement of material.
4. Waiting	Waiting on upstream activity.
5. Inventory	Storage and handling of material that cannot be used immediately.
6. Motion	Movement of people or materials without adding value.
7. Processing	Performing unnecessary tasks.

Figure 2.1
Lean production
concepts
*Adapted from Toyota
Production System
house*

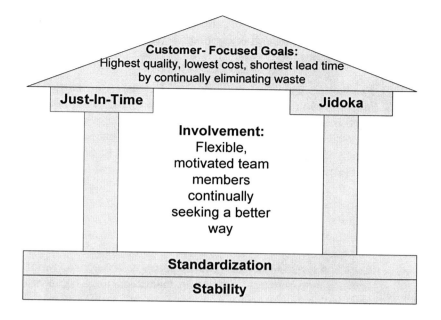

2.2.2.1 Stability and standardization

The foundations of lean production are stability and standardization. They bring order to the chaos that is an inherent part of production systems. Stability and standardization are mutually supportive. Standardization provides stability to the workplace and stability is required for a production system to meet standard (expected) levels of performance. A conceptual starting point is visual management or 5S system. Visual management organizes the workplace so that no problems are hidden – it is readily apparent what should be done, what is being done, and whether there is a deviation. This creates a self-managing work environment that is self-explaining, self-ordering and self-improving [13]. The 5Ss include:

- *Sort* through all items in the workplace and dispose of rarely used items.
- *Straighten* by organizing and locating the remaining items so that they are easy to find, locate and return.
- *Shine* by cleaning the work area and maintaining the equipment so that it is ready for use.
- *Standardize* by developing procedures to formalize and maintain the first three Ss.
- *Sustain* by creating the organizational environment that will sustain commitment – awareness, time, structure, support, rewards/recognition, satisfaction and excitement.

The first three Ss are useful when configuring the factory. Unused items (materials, tools/equipment, floor space) should be disposed. They create problems – congesting the workplace making it difficult to work and move; making it hard to find the right tools and materials and easier to use the wrong tools and materials; occupying valuable floor space and storage equipment, raising costs, lowering capacity and making it harder to arrange the workplace efficiently; and hiding other problems. Frequently used items should be organized and located so that they are easy to find, locate and return. Materials and equipment should be located to facilitate cleaning, inspection and maintenance.

Production leveling (or "heijunka" in lean terminology) also contributes to stability by accumulating orders over some period and distributing the volume and mix evenly for each production day. Leveling the peaks and valleys of customer demand allows the production system to produce at the same pace every day, eliminating unevenness in production levels (or "mura" in lean terminology), overburden to people and equipment (or "muri" in lean terminology), and waste (or "muda" in lean terminology).

Standardization refers to the intentional design of both products and processes to achieve greater commonality and repeatability. Enhancing product commonality from the manufacturing perspective need not be accomplished by unduly limiting customer choice. The automotive and electronics industries have been highly successful in exploiting common component parts, sub-systems, modules, and design features to limit design-induced variation in manufacturing, while offering wide customer choice. Standard manufacturing processes are the safest, easiest and most productive ways of doing the job. They are also the baseline from which improvements are continually made, and thus are transitory. A standardized process defines the tasks that must be performed, the materials (including inventory size) that are used, the tools/equipment that are used, the layout of the immediate workplace, the time required to perform each task, and the sequence in which tasks are performed.

2.2.2.2 Just-in-time and jidoka

The two supporting pillars for lean production are just-in-time (JIT) and jidoka. Together, they provide the customer with what is valued – the right product, with the right quality, at the right time, in the right quantity. JIT seeks to create this value by providing a continuous flow of work on the product, which simultaneously eliminates all waste and provides the quickest response to the customer. Ideally, the product is processed through an activity and then passed immediately to the next activity where it begins processing without delay. The ideal continuous flow production system has the following features:

- Single piece flow – The product is processed through each activity and then passed to the next activity individually, in a production lot size of one. If the product were part of a larger lot, processing at the next activity would be delayed: 1) until the remainder of the lot was finished before being passed to the next activity and 2) after becoming available at the next activity, until the previous lot was finished and then awaiting its turn in its own lot. This would extend overall production cycle time, increase work in process (WIP) inventory and delay detection of potential defects.

- Quick setup – To mitigate the increased number of setups (and teardowns) associated with single piece flow, setup must be performed quickly. Single Minute Exchange of Die (SMED) techniques are used to reduce setup times. Simplifying critical setup tasks and differentiating external setup tasks (those tasks that can be performed before setup starts, in parallel with the previous production run) speed setup.

- Synchronized flow – Every product in the production process completes its current activity and moves to the next activity at the same time. This allows the production system to fully capitalize on the benefits of single piece flow. It also establishes a highly visible work pace that promotes efficiency and readily exposes problems that require resolution.
- TAKT time – TAKT time is the rate of demand expressed as average cycle time per unit of demand. For example, a TAKT time of two hours indicates that, on the average, one unit is sold for every two hours of production time. The TAKT time is used as the synchronized flow cycle time for each activity in the production process. It is also called the "drumbeat" or "heartbeat" that paces flow throughout the production process.
- Balanced workload – The workload is balanced among workers (and teams of workers) so that all activities are completed within the TAKT time. This satisfies demand while equalizing work assignments and maximizing efficiency and capacity.
- Proximity – Workers and equipment that perform sequential activities are located close together so that the continuous flow of value-adding processes is not interrupted. The production process is often arranged as a straight, L-shaped, or U-shaped line. U-shaped lines often provide the best combination of operating efficiency, space utilization and expansion flexibility.
- Stability and standardization – The production process is stable, demand has been leveled and the product and process has been standardized as described earlier.
- Technology – Only reliable, thoroughly tested technology that serves workers and processes are used. Technology must be flexible to accommodate continuous process improvement as well as larger scale business changes.

Pure continuous flow is not always practical or even possible. Several scenarios can disrupt continuous flow. An upstream supplier activity may produce parts in batches far more efficiently than individual parts, even after searching for a simpler alternative process and reducing setup times. Product completing a supplier activity may require time to dry, cool, cure, etc. Elemental component parts that are produced by a supplier activity may be used in a number of downstream customer activities that are scattered throughout the plant, and it is not practical to: 1) co-locate all of the customer activities around the supplier activity, 2) duplicate the supplier activity for each customer activity, or 3) transport single parts throughout the plant to facilitate continuous flow. Each of these scenarios may necessitate a lot size greater than one and/or some form of queueing.

Perhaps the most common obstacle to continuous flow is process time variation. Variation may be random, emanate from product design variation (model mix and/or unique customization) or result from wasteful inefficiencies or delays (rework due to poor workmanship or damage, material unavailability or poor quality, design questions or errors, change orders, equipment breakdowns, tool malfunctions). If variation causes process time to exceed TAKT time, then there is a risk of delaying downstream activities. If there is no queueing space between activities, then there is also a risk that the delay will block upstream activities. Lack of queueing space may also cause blockage when process time is less than TAKT time – preventing completed product from promptly exiting the activity. When excessive process time variation afflicts a bottleneck activity (where the average process time approaches or exceeds TAKT time), the disruption of continuous flow is more severe. At a bottleneck activity lost time cannot be recovered, resulting in potential capacity loss unless extraordinary measures are taken. A variety of remedies can be employed to increase capacity and alleviate the bottleneck [14]:

- Eliminate wasteful inefficiencies and delays – design errors or questions due to incomplete design documentation; change orders; material unavailability due to poor incoming quality or replenishment failures; rework due to poor workmanship or damage; equipment breakdowns or tool malfunctions.
- Rationalize the activity to increase efficiency.
- Increase production hours – work during breaks and lunch or add overtime.
- Add resources – workers, equipment, floor space.
- Establish queues around bottleneck – inbound queue prevents starvation; outbound queue prevents blockage and downstream starvation.
- Unbalance line capacity – prevent starvation by providing excess capacity at upstream activities.

Liker [11] suggests that when it is not possible to create pure continuous flow, the next best approach is to establish a controlled inventory between affected activities and "pull" from the inventory. Rother and Shook [15] summarize: "Flow where you can, pull where you must". The basic strategy of a pull system is to avoid overproduction and the resulting uncontrolled growth in inventory by linking an upstream supplier production activity to downstream customer demand. The customer is the entity that uses the product – either the actual purchasing customer (external) or a downstream customer production activity (internal). The pull system is characterized by two elements, supermarket and kanban. Ohno [8] created the concept of supermarket or small store as a compromise between the ideal of continuous flow and a traditional, less controlled "push" system. The inventory level for each item in the supermarket is set to the minimum required to provide smooth flow. When a specified quantity of the item is withdrawn from the supermarket by the customer, a visual signal (or "kanban" in lean terminology) such as a card or empty container is returned to the supplier activity to initiate production of the specified quantity.

Building-in quality (or "jidoka" in lean terminology) is the second supporting pillar of lean production. Jidoka implies intelligent workers and machines delivering perfect first-time quality. Jidoka involves developing processes with high capability (producing few defects) and containing the defects that do occur by automatically identifying them and taking quick countermeasures. The concept of using simple, low-cost devices that either detect defects before they occur or identify them immediately and stop the line is called "poka-yoke" in lean terminology. These devices, typically contact or non-contact sensors, detect deviations in the part or in work methods.

Jidoka is facilitated by single piece flow. Each worker is also an inspector who identifies and initiates repair on incoming defects before passing them to the next worker. Defect detection is faster in single piece flow. The worker responsible for the defect is notified immediately, both to repair the immediate defect and to resolve its root cause to prevent recurrence. When necessary, workers can stop the line when problems are encountered. As important as it is to maintain continuous flow, it is more important to identify and resolve problems as they occur, particularly those that threaten to disrupt continuous flow on an ongoing basis. A visual control device, such as a status light, that communicates when a problem has arisen, is called an "andon" in lean terminology.

2.2.2.3 Employee involvement

At the core of lean production is an empowered workforce, focused continually on seeking a better way. The objective is to nurture and utilize the vast potential of production workers to improve their own operations while simultaneously improving the organization's prospects for long-term success. Employee empowerment starts with operational responsibility and control of his/her own assignment – producing product within TAKT time and assuring that it meets quality specifications. The worker is empowered to stop the line when necessary to identify and resolve critical problems jeopardizing this assignment. Beyond this immediate assignment, workers are responsible for continually improving their workplace, including participation in 5S teams and development of process standards. At the broadest level, a worker may be called to serve as a member of a continuous improvement team that is formed to address cross-cutting issues that extend beyond his/her workplace. Examples include efforts to reduce supermarket inventory levels or to eliminate process inefficiencies/delays.

Continuous improvement teams are a successful vehicle for organizing worker involvement to implement lean production principles. Continuous improvement (or "kaizen" in lean terminology) can be defined [16] as "the planned, organized and systematic process of ongoing, incremental and company-wide change of existing practices aimed at improving company performance." In contrast to scientific management approaches that split employees into "thinkers" and "doers," kaizen assumes that all employees can make a contribution to problem solving and innovation [17]. Kaizen eliminates waste by empowering employees with the responsibility, time, tools and methodologies to uncover areas for improvement and plan and implement change. Kaizen is team-based and should involve workers from different levels of the organization. The kaizen blitz takes the same improvement philosophy and applies it in a brief, but intense attack on production waste and inefficiency [18]. Both kaizen methods follow the structured approach shown in Figure 2.2.

Figure 2.2
Lean production implementation
methodology

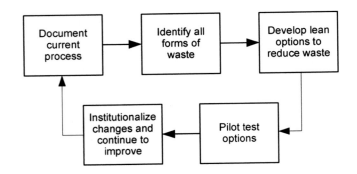

Lean production methods focus on the value stream, the set of activities used to create a product or service from raw material until it reaches the customer [10]. The first step involves the development of a value stream map (VSM) that documents the flow of product and information between process activities [15]. Critical performance metrics (quality, cycle time, productivity, inventory) are documented for each activity. Waste in all its forms is exposed as the current process is observed, documented and analyzed. Non-value added activities are discovered. Delays and inefficiencies are revealed. When waste is identified, potential process improvements are developed using lean principles. Selected lean improvements are pilot tested in the process and fine-tuned to optimize impact. As the successful changes are institutionalized, the continuous improvement process is repeated in a never-ending cycle. VSM can be performed at different levels of the organization: specific process activity, door-to-door within the plant, across the enterprise, or across organizations to suppliers and customers.

2.2.2.4 Lean construction

Modular production bears a striking resemblance to site-built housing construction, with several important differences: 1) it takes place inside a factory on a moving production line and 2) construction crews are a dedicated resource, expediting the "parade of trades". Given the similarity of modular production to site-built construction, a review of lean applications in construction is instructive. Koskela [19] uses lean production principles to derive the philosophy of lean construction. Lean construction begins with a lean project plan, the set of construction activities organized so that they can be completed with the highest quality, lowest cost, and shortest lead time. Project organization starts with a list of all activities required to complete the project, the time required for each activity, and the precedences between activities.

Using these values, a technique such as the Critical Path Method [20] is used to determine the critical path – the sequence of activities that form the lengthiest path (greatest overall time) through the project. This is the shortest time required to complete the project. Any delay of an activity on the critical path directly impacts the completion date. The technique also determines the earliest and latest that other activities can start and finish without impacting the completion date. The results allow managers to reorganize the project to shorten the planned critical path by pruning critical path activities, "fast tracking" (performing more activities in parallel), and/or "crashing the critical path" (shortening the time of critical path activities by adding resources). The results also allow the manager to prioritize activities for management attention during project implementation.

When the lean project plan has been developed, it must be implemented. Like lean production, a fundamental principle of lean construction is maintaining a continuous flow of value-adding work on the project. However, since the construction project itself cannot move, construction subcontractors must move continuously through the product. Ballard and Howell [21] state that achieving reliable workflow is possible only when sources of variability are controlled. They identify the quality of weekly crew assignments as a key driver of variability, with quality assignments shielding downstream production crews from work flow uncertainty. System performance is measured by the percentage of the planned work assignments that were successfully completed. Ballard and Howell [22] suggest that remaining workflow variability can be mitigated through the use of plan buffers, surge piles and flexible capacity. Plan buffers refer to a backlog of reliable assignments for crews. Surge piles take the form of raw and processed materials ready for use in these assignments. Flexible capacity refers to intentional underutilization of a crew or the flexible use of cross-trained workers.

2.2.3 Mass customization

As homebuyers become more sophisticated, modular producers are routinely customizing standard models and increasingly building one-of-a-kind custom homes, multi-family condominiums and apartments. Modular manufacturers may actually be producing a more customized product than some production builders, who have been intentionally narrowing design options. Using data from the SBRA benchmarking effort [4], Nahmens [23,24] found that producers offering increased product choice are likely to suffer poorer labor productivity, greater inventory, higher production costs, more quality issues, less satisfied homebuyers, and lower space efficiency. In general, operational performance deteriorated with an increase in product choice. Deterioration stems from an increase in the number of different materials that must be handled and variability in the production process associated with varying design features: materials, number of parts (windows, interior walls), dimensions (module length), and design features (bumpouts, roof pitch, dormers). The number of different materials affects staging and handling in production areas as well as in backup storage areas. Production process variation changes the basic process methods/tasks, the tools/equipment, the physical quantity of work, and the resulting cycle time. This variation disrupts continuous flow and exacerbates potential bottlenecks. Modular producers have identified a number of activities that can bottleneck production. Roof and wall framing were cited most often as bottlenecks, followed by the line configuration and drywall finishing. Mullens [25] has conjectured that multiple "floating bottlenecks" are a primary culprit in reducing plant capacity. A floating bottleneck is a bottleneck that shifts between activities, depending on product mix [26]. From the worker's perspective, production process variation increases the skill level requirements and disrupts the worker's steady, sustainable pace, causing the worker to hurry up and wait.

Frustration and exhaustion are the natural consequences and safety, quality and job satisfaction are likely victims.

Mass customization is the ability to design and manufacture customized products at mass production efficiency and speed. Three mass customization strategies have been proposed for the industrialized housing industry [23,24]: modular product architecture, postponement of customization, and flexible production processes. The strategies are most effective when applied simultaneously, using innovative new product development methodologies like concurrent engineering [27,28]. It should be noted that successful mass customization still requires the modular producer to limit customer choice. However, when coupled with excellence in architectural design and thorough market research, these strategies should yield the widest practical range of choice with the least impact on production. The strategies are summarized in the following sections.

2.2.3.1 Modular product architecture

Mass customization can be achieved by employing flexible modular product architecture, such as a product platform design incorporating common modular components [28]. Note that modular architecture in this context refers to product structure – products composed of mix-and-match modular components and the coupling between these components. Flexible modular product architecture can contribute to a product's market appeal (look/feel, functional performance, range of choice) and its ease of production. For example, Dell utilizes highly modular product architecture to drive their successful personal computer (PC) business. Each functional element of the PC is designed as a plug compatible, substitutable module available in a limited number of style, performance and cost options. Modules are procured when needed from a worldwide supply chain. Customers configure their own PC by selecting from available options, and Dell produces each customer order by assembling the selected modular components.

From a production perspective, modular product architecture offers the following advantages:
- Processes used to sub-assemble modular components have a smaller scope and are more focused. Tools/equipment, methods and worker skills are specialized. Therefore, the subassembly process is more efficient and reliable. Efficiency and quality is maximized when the subassembly process is limited to a common modular component or a small family of related components.
- When there is variation in the modular component, the smaller scope of the subassembly process can still be advantageous. While the narrower scope of a subassembly process can make it more sensitive to product variation, it often makes it easier to accommodate this variation. For example, the smaller scope contributes to greater commonality among subassemblies, allowing more process specialization and build-to-stock (a postponement mass customization strategy) versus build-to-order production. The smaller scale of modular components also makes it easier to add/rearrange production equipment and add queueing.

- Subassembly processes can be performed at the same time as (parallel with) higher level work. If coupling of the subassembly to the higher level assembly is kept simple, this reduces the volume, complexity, duration and variation of the higher level assembly work. If the higher level assembly is on the critical path, this correspondingly reduces overall production cycle time, eases bottlenecks and facilitates an increase in capacity.
- Modular components can be outsourced to vendor specialists. This allows access to advanced material and production technologies not otherwise available to the producer due to low volume and limited capital resources. Outsourced modular components can offer better functional performance, higher quality and/or lower cost. Advanced vendor production technologies such as flexible manufacturing may also be more capable of accommodating variation in modular components.

2.2.3.2 Postponement

In every supply chain, there is a differentiation point – the point where a specific product becomes associated with a specific customer order. Operations upstream of this point are buffered from unpredictable fluctuations in customer demand. They are highly standardized to reflect the more standard product. Production is stable, flowing continuously in a build-to-stock mode, filling work-in-process (WIP) inventories of partially completed goods. Operations downstream of the differentiation point are driven by specific customer orders and function in a build-to-order mode. They reflect all the variation in the orders and are not so standardized. Flow is not so continuous. Postponing the differentiation point can add stability to the overall supply chain. Toyota Home uses an extreme form of postponement, producing small standard modules that are shipped to the construction site and assembled to create a custom home [29]. Since the differentiation point occurs after distribution, specific customer orders have no impact on the production process. The highly modular product architecture used by Dell allows computer customization to be postponed to final assembly. In final assembly, customer specified modular components are added to the basic product platform. These examples show how modular product architecture is used to enable a postponement strategy for mass customization.

The conventional wood frame building system utilized by most producers is not very conducive to postponement. Customer specified product variation is integral to most major subassemblies and cannot be readily postponed. However, reducing the scale of major subassembly (discussed above) increases the likelihood of more common elements and might provide similar advantages.

2.2.3.3 Flexible production process

The production process can be configured so that it is less sensitive to product variation dictated by the customer. In general, the process' ability to accommodate product variation is enhanced by 1) reducing the cycle time variation that results from product variation and 2) reducing the negative impacts that result from any remaining cycle time variation. This can be accomplished in many ways including:

- Elemental process selection – Consider elemental processes that are not only safe, reliable, and highly efficient, but that contain cycle time variation caused by product variation.
- Process organization – Organize elemental processes around a modular product architecture. Produce each subassembly independently and then flow to the higher assembly level.
- Process refinement – Use lean production techniques to eliminate waste and make the process less sensitive to product variation. For example [24]:
  o Eliminate wasteful inefficiencies and delays that add cycle time variation: rework due to poor workmanship or damage, material unavailability or poor quality, design questions or errors, change orders, equipment breakdowns, tool malfunctions.
  o Standardize and rationalize elemental processes, particularly bottlenecks that are affected by product choice. Develop common methods, equipment and tools that make the process both quick and insensitive to cycle time variation. Note that quickness itself makes the process less sensitive to variation.
  o Move equipment and materials closer together. Utilize straight line, L or U-shaped product flows. This reduces travel waste such as excessive travel time, congestion delay, and related damage. It also reduces the cycle time variation associated with the number of trips required to get material for different product configurations.
  o Flow continuously. When production flow needs to be disconnected due to cycle time variation, use limited queues with kanbans to drive production. When product choice results in many components, consider pulling materials in built-to-order kits, instead of unique part numbers. This can control inventories in the workplace, even as product choice increases.
  o Schedule production so that product variation is spread out over time. Level the peaks and valleys of customer demand over the week so that the production process produces at the same pace every day. Daily production should be sequenced to prevent consecutive "problem" modules from successively choking the same bottleneck. Try to anticipate potential bottlenecks and add resources such as cross-trained utility workers.
- Worker mobility – Enable workers to move downstream to complete their work or to move upstream to begin their work early. This allows workers to absorb variations in cycle time without disrupting flow. Worker mobility can extend beyond their assigned activity. When finishing early, workers might help nearby colleagues struggling to complete work on a different activity. Worker mobility can be extended even further, creating a general category of worker (often termed flex or utility worker) that is responsible for moving throughout the plant and assisting any work team needing assistance. Support systems can enhance worker mobility: cross-training gives workers needed skills; incentives encourage workers; and information systems and supervision provide real-time guidance.

- "Factories within a factory" – Produce diverse or customized products on independent lines. Each line will have less product variation with fewer, more easily controlled bottlenecks.
- Flexible manufacturing – A flexible manufacturing system (FMS) typically consists of computer numerically controlled (CNC) machines, linked by an automated material handling and storage system, all under the control of an integrated computer control system. An FMS typically has flexibility in four dimensions: volume, manufacturing processes, product mix, and delivery. FMS technology is capital intensive.

2.3 REFERENCES

1. Systems Building Research Alliance, *Applying Lean Home Building Practices Beyond the Plant: Final Report*, New York State Energy Research and Development Authority, Albany, NY, September, 2010.
2. Mullens, M. and M. Kelley, "Lean Homebuilding Using Modular Technology," *Housing and Society*, 31(1), 41-54, 2004.
3. Manufactured Housing Research Alliance, *Develop Innovations in Manufacturing Processes through Lean Production Methods*, U.S. Department of Housing and Urban Development, Affordable Housing Research and Technology Division, Washington, D.C., October, 2005.
4. NAHB Research Center, Inc., *Factory and Site-Built Housing: A Comparison for the 21st Century*, U.S. Department of Housing and Urban Development, Office of Policy Development and Research, Washington, D.C., October, 1998.
5. Manufactured Housing Research Alliance, *Pilot Study: Applying Lean to Factory Homebuilding*, U.S. Department of Housing and Urban Development, Office of Policy Development and Research, Washington, D.C., July, 2007.
6. Elshennawy, A., M. Mullens, I. Nahmens, "Quality Improvement in the Modular Housing Industry" *Industrial Engineering Research '02 Conference Proceedings*, Orlando, May, 2002.
7. Mullens, M., "Quality Systems and Improved Plant Design," presentation given to the Quality Modular Building Task Force, Salem, MA, October 2001, http://www.housingconstructabilitylab.com/pages/qual%20pres%20[Compatibility%20Mode].pdf, May, 2009.
8. Ohno, T., *Toyota Production System*, Productivity Press, New York, 1988.
9. National Association of Home Builders (NAHB) Research Center, *Lean Construction*, Upper Marlboro, MD, 1999.
10. Womack, J. and Jones, D., *Lean Thinking: Banish Waste and Create Wealth in Your Corporation*, Simon & Schuster, New York, 1996.
11. Liker, J. *The Toyota Way: 14 Management Principles from the World's Greatest Manufacturer*. McGraw-Hill, New York, 2004.
12. Cusumano, Michael and Nobeoka, Kentaro, *Thinking Beyond Lean*, The Free Press, New York, NY, 1998.
13. Grief, Michel, *The Visual Factory*, Productivity Press, Portland, OR, 1991.
14. Goldratt, E., *The Goal*, North River Press, Great Barrington, MA, 1992.
15. Rother, M., and J. Shook, *Learning to See*, The Lean Enterprise Institute, Brookline MA, 1999.

16. Boer, H., Berger, A., Chapman, R. and Gertsen, F. (Eds.). *CI Changes: from Suggestion Box to Organizational Learning – Continuous Improvement in Europe and Australia*, Ashgate, Aldershot, 2000.

17. Bessant, J., Caffyn, S., and Gallagher, M. "An Evolutionary Model of Continuous Improvement Behaviour," *Technovation*, 21, 67-77, 2001.

18. Laraia, A., Moody, P., Hall, R. *The Kaizen Blitz: Accelerating Breakthroughs in Productivity and Performance*, Jossey-Bass, San Francisco, 1999.

19. Koskela, L., "Lean Production in Construction", *Proceedings of the 10th International Symposium of Automation and Robotics in Construction*, pp. 47-54, Houston, TX, May 24-26, 1993.

20. Woolf, Murray B. *Faster Construction Projects with CPM Scheduling*. McGraw Hill, New York, 2007.

21. Ballard, G. and Howell, G., "Implementing Lean Construction: Stabilizing Work Flow", *Proceedings of the 2nd Annual Meeting of the International Group for Lean Construction*, Santiago, Chile, 1994.

22. Ballard, G. and Howell, G., "Shielding Production: Essential Step in Production Control", *Journal of Construction Engineering and Management*, 124(1), 11–17, 1998.

23. Nahmens, I. and M. Mullens, "The Impact of Product Choice on Lean Homebuilding," *Construction Innovation: Information, Process and Management*, 9(1), 84-100, March 2009.

24. Nahmens, I. *Mass Customization Strategies and Their Relationship to Lean Production in the Homebuilding Industry. Ph.D. Dissertation*, University of Central Florida, Orlando, FL, August, 2007.

25. Mullens, M., "Production Flow and Shop Floor Control: Structuring the Modular Factory for Custom Homebuilding" *Proceedings of the NSF Housing Research Agenda Workshop*, Feb. 12-14, 2004, Orlando, FL. Eds. Syal, M., Mullens, M., and Hastak, M. Vol 2, pp. Focus Group 1.

26. Hopp, W. and Spearmann, M., *Factory Physics: Foundations of Manufacturing Management*, 2nd Edition, Irwin/McGraw Hill, New York, 2001.

27. Rosenblatt, A. and Watson, G., "Concurrent Engineering," *IEEE Spectrum*, pp. 22-37, July, 1991.

28. Ulrich, K. and Eppinger, S., *Product Design and Development*, McGraw Hill Inc., New York, 1995.

29. Barlow, J., Childerhouse, P., Gann, D., Hong-Minh, S., Naim, M., and Ozaki, R. "Choice and Delivery in Housebuilding: Lessons from Japan for UK Housebuilders". *Building Research and Information*. 31(2), 134-145, 2003.

CHAPTER 3
MODULAR HOMEBUILDING COMPONENTS AND PRODUCTION PROCESSES

Chapter 3 describes the primary factory-built components used in modular homebuilding and the processes used to produce them.

3.1 THE MODULE

The module is the basic building component in modular homebuilding and is the primary product of the modular factory. Each module includes the floor, walls and roof/ceiling with plumbing and electrical systems installed and interior and exterior finishes applied. Each standard home design has a unique set of associated module designs. Even when two homebuyers order homes based on the same standard home design, module designs are rarely identical. This is the result of custom changes demanded by the homebuyer. Gianino [1] describes how the character of a standard home design can be creatively transformed by changing room sizes and shapes, adding/deleting rooms, adding a bump-out or a wing, increasing roof pitch, changing from a gable to a hip roof, adding dormers, changing type and location of windows, and adding other major design elements such as a porch, garage, or fireplace. The transformation can be further enhanced by substituting custom finish elements such as exterior siding, shingles, wood/tile flooring and custom kitchen/bath components.

Findings from the SBRA benchmark survey [2] provide insight about the home designs offered and sold by modular producers. Modular producers offer an average of 82 standard home models (Figure 3.1). Of these offerings, an average of only 35 models make up 90% of sales (Figure 3.2). Almost all modular homes are customized. Modular producers report that 10% of homes are not customized, 34% have minor customization, 40% have extensive customization, and 16% are totally unique custom homes (Figure 3.3). The customization profile varies widely by producer, with some encouraging customization and others discouraging it. Almost all homes require multiple modules, with more than half requiring two modules (Figure 3.4). The average home requires 2.7 modules. Thirty-four percent of homes are multi-story (Figure 3.5). To provide higher roof pitches, most modular factories only produce homes with hinged roofs. Eighty-five percent of homes have hinged roofs (Figure 3.6). Almost all modular producers install finished drywall throughout their homes [2].

3.1.1 Limitations on Module Size

Module size is limited by highway transport considerations. Typical module widths offered by modular producers include 12', 13' and 13'9". Some producers can also produce wider modules such as 14'9" and 15'9". However, these are subject to more restrictive route and schedule limitations, highway fees and escort requirements. Minor variations from typical widths can be accommodated by many producers. Hinged eaves allow modular producers to maximize module width while accommodating a generous eave overhang.

Figure 3.1
Profile of total standard home models for modular producers [2]

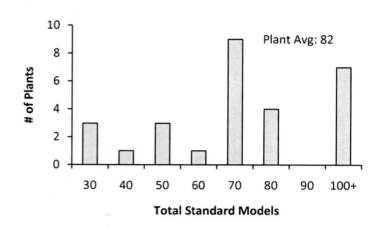

Figure 3.2
Total standard home models comprising 90% of sales for modular producers [2]

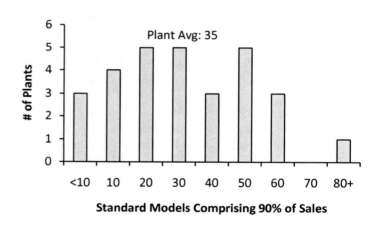

Figure 3.3
Profile of customization of homes sold by modular producers [2]

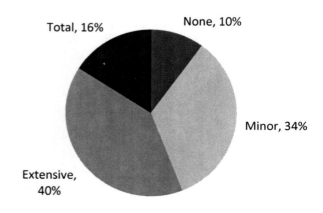

Figure 3.4
Profile of number of
modules for homes sold by
modular producers [2]

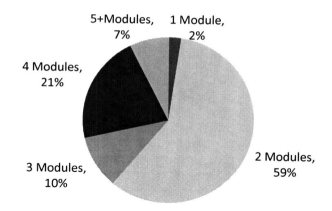

Figure 3.5
Profile of multi-story homes
for modular producers [2]

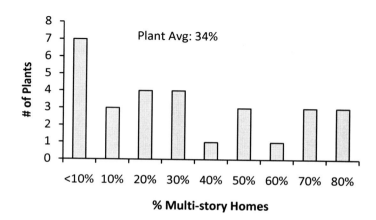

Figure 3.6 Profile of homes
with hinged roofs for
modular producers [2]

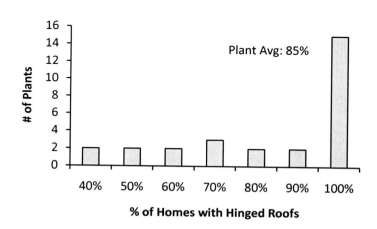

Most modular producers limit module length to 60'. Although some states allow longer modules to be transported, they are subject to more restrictive route and schedule limitations, highway fees and escort requirements. Module length typically averages 45–50'. Although shorter modules can be built, they can reduce line capacity. Modular producers often build and transport multiple smaller modules of similar width as a single large module.

Module height, when loaded on a transport carrier, is typically limited to 13'6" to clear overhead roadway obstructions. The bed of a typical carrier is 2'6" high, limiting actual module height to 11'. While this may be satisfactory for modules without a roof (the first floor of a two-story home), most modules containing a roof cannot meet this limitation without design modifications. The specific design utilized depends on the height of the wall, the width of the module and the pitch of the roof. Incorporating one or two hinged elements in the roof design will usually allow the design to meet the limitation. If the module is still too high, it may be produced without the roof, and the roof added at the construction site. The roof may be site-built, assembled from factory-built panels, or set independently as a custom factory-built roof module.

3.2 COMPONENTS AND PROCESSES

The modular factory performs a wide range of production tasks involving cutting, assembling and finishing building materials. These tasks are described in the following sections, including typical building materials, major process equipment, and production methods. Power hand tools such as nail guns, screw guns, circular saws and routers are widely used for most tasks. To simplify the task descriptions, tasks are organized into the activities shown in Table 3.1. This organization assumes that the major structural elements (floor, walls, and roof/ceiling) are built as independent subassemblies, instead of built-in-place on the module. It also assumes that a small number of minor subassemblies are built: window/door openings, various plumbing subassemblies, interior doors and various roof components (knee walls, folding ridge panels, folding soffits). Later chapters address how different organizations of the product and process can enhance production performance.

It should be noted that some tasks that might be performed in the factory are routinely performed on site. For example, HVAC system installation (other than hot water heating) is routinely performed on site. This is largely due to the desire of the builder and the homeowner to give complete turnkey responsibility for the HVAC system to a local supplier. Although duct systems can be installed in the factory, it is usually just as convenient to install in the field – first-floor ducting can be installed from the basement or crawl space and second-floor ducting can be installed from the attic. Furthermore, it is seldom possible to integrate the entire HVAC system into the factory-built modules. Condensers and heat pumps are typically located on a pad outside the home and air handlers are often located in a site-built garage, attic or basement.

3.2.1 Cut framing components (mill)

Ideally, a modular producer purchases all raw materials pre-cut to the exact size needed in the module design. This is sometimes possible for lumber used in high volume applications with little design variation, such as wall studs. In general, however, volume purchasing discounts and home-to-home design variation prevent the producer from purchasing some high volume components pre-cut to the correct size. To reduce the workload and cycle time in the critical framing areas, a mill area equipped with high capacity component saws, chop saws, and panel saws is used to cut some components to size before they are delivered to the framing areas. Typical lumber components include floor joists, ceiling joists, wall studs, window framing components, roof components, and blocking. Sheathing components for the roof and wall may also be cut to size in the mill. Even when purchasing pre-cut lumber is economically feasible, some producers chose to cut all components to size to assure more precise dimensions and eliminate ends damaged in shipping and handling.

Table 3.1 Production Activity List for Typical Modular Factory

	Activity		Activity
1	Cut framing components (Mill)	24	Sheath walls
2	Build floor	25	Install windows and exterior doors
3	Build window/door opening subassemblies	26	Install siding and trim
4	Build partition walls	27	Hang drywall on walls
5	Build side walls	28	Tape and mud drywall
6	Build end walls	29	Sand and paint
7	Build marriage wall	30	Install cabinets and vanities
8	Set partition walls	31	Fabricate and install kitchen countertops
9	Set exterior and marriage walls	32	Build finish plumbing subassemblies
10	Install rough electric in walls	33	Install finish plumbing
11	Build plumbing subassemblies	34	Install finish electric
12	Install rough plumbing in wall and tubs	35	Build interior door subassemblies
13	Build subassemblies for roof	36	Install interior doors
14	Build roof/ceiling	37	Install molding
15	Set roof	38	Install miscellaneous finish items
16	Install rough electric in roof	39	Install flooring
17	Install rough plumbing in roof	40	Load shiploose
18	Insulate roof	41	Factory touch-up
19	Sheath and install subassemblies for roof	42	Install plumbing in floor
20	Shingle roof	43	Load module on carrier
21	Prep/drop roof and wrap for shipment	44	Final wrap and prep for shipment
22	Install fascia and soffit	45	Build major shiploose subassemblies
23	Insulate walls		

3.2.2 Build floor

A single floor is produced for each module. Floor dimensions are equal to those of the module footprint. The floor is built on a specialized framing jig table (Figures 3.7–3.8). The table may have a solid top or simply be a structural frame. More sophisticated framing jigs may incorporate a sliding/folding/rolling frame that adjusts for differing floor widths, adjustable positioning jigs for each joist, and a rail-mounted nail gun on each side. A chop saw with an extended length bed is provided to cut rim joists and other components to size. Primary materials used include framing lumber (2"x8", 2"x10" and 2"x12" full length rails and pre-cut joists), floor sheathing (4'x8'x3/4" plywood or OSB), underlayment (1/4" or ½" plywood or luan) and vinyl floor coverings. The basic process includes:

Framing
1. Cut end and side rim joists to size using chop saw and mark joist placement on inner side rim joists (16" on center).
2. Build box-out subassembly(s) if required.
3. Place joists and box-out subassembly(s) on table.
4. Place inner rim joists on table and attach using nails.
5. Position each joist along rim joists and attach using nails. Install using joist hangers if specified.
6. Position blocking between joists down center of frame and attach using nails.
7. Insulate along perimeter if required using batt insulation and netting to hold in place.
8. Move outer rim joists to table, position for joint overlap and attach using nails.
9. Install gang nails at all butt seams on marriage side.

Sheathing
1. Square frame and lag to table to prevent movement.
2. Apply continuous bead of adhesive on joists.
3. Move sheathing to table, position and attach using nails or screws.
4. Route excess sheathing around perimeter of floor and around specified openings.

Underlayment (if required)
1. Cut underlayment to size using panel saw.
2. Position underlayment on floor and attach using nails.
3. Sand underlayment, ensuring all seams are flush.
4. Sweep and blow debris from floor.

Vinyl flooring (if required)
1. Pull vinyl off rack and cut to size.
2. Position vinyl over underlayment, stretch and attach around perimeter and openings using staples.

Figure 3.7
Framing table used to
build floors

Figure 3.8
Framing table used to
build floors

3.2.3 Build window and door opening subassemblies

Framing for window and door openings used in the exterior walls are built on a small framing jig table (Figure 3.9). A chop saw is provided in the area to cut framing components to size. Materials used include: framing lumber (2"x4" and 2"x6" full-length studs and pre-cut components) and wall sheathing for headers (½" plywood or OSB). The basic process includes:
1. Build header subassemblies.
2. Cut components as required and mark for assembly.
3. Place components and header subassembly on table.
4. Attach using nails.

Figure 3.9
Framing table used to build window
and exterior door openings

3.2.4 Build partition walls

The number of partition (interior) walls that are produced for each module varies greatly, often ranging from 3–14 walls per module. Partition wall length may vary from the depth of a closet to the width of the module. Partition walls are built on several types of framing jigs, either horizontal tables (Figures 3.10–3.11) or vertical A-frames (Figures 3.12–3.13). The A-frame jig provides two framing surfaces (one on each side), with only a modest increase in floor space. More sophisticated framing tables may incorporate a sliding/folding/rolling frame that adjusts for differing wall heights, adjustable positioning jigs for each stud and a rail-mounted nail gun on each side. More sophisticated A-frame jigs may also incorporate adjustable positioning jigs for each stud. When using a longer framing jig, partition walls are often batched and produced at the same time to maximize jig utilization. A chop saw with an extended length bed is provided to cut top and bottom plates and other components to size. Primary materials used include precut framing lumber for studs (2"x4" and 2"x6" in 8' and 9' lengths), framing lumber for plates (2"x4" and 2"x6" in 14' lengths), and drywall (1/2" drywall in 4'x8', 4'x14' and 54"x14' dimensions). The basic processes are similar to those of the sidewall, which are described in the following section.

Figure 3.10
Framing table used to build
partition walls

Figure 3.11
Framing tables used to build partition and marriage walls (left) and side walls (right)

Figure 3.12
A-frame framing jig used to build partition walls

Figure 3.13
A-frame framing jig (background) used to build walls

3.2.5 Build side wall

A single side wall (long exterior wall) is produced for each module. The length of the side wall is equal to the length of the module. Standard wall height is 8'. Most producers also offer an optional 9' height, and some factories will provide variations. The wall is built on a long framing jig similar to that used to build partition walls, either a horizontal table (Figures 3.14–3.16) or A-frame (Figure 3.17). A chop saw with an extended length bed is provided to cut top and bottom plates and other components to size. Primary materials used include framing lumber for precut studs (2"x4" and 2"x6" in 8' and 9' lengths), framing lumber for plates (2"x4" and 2"x6" in 14' lengths), pre-built window/door opening subassemblies, and drywall (1/2" drywall in 4'x8', 4'x14' and 54"x14' dimensions). The basic process includes:

Framing
1. Measure length of top and bottom plates and mark locations of studs (typically 16" on center) and window/door opening subassemblies.
2. Using a chop saw, cut top and bottom plates to size.
3. Place bottom plates, window/door opening subassemblies, studs and first top plates on framing jig.
4. Position each subassembly and stud along plates and attach using nails.
5. Splice bottom plate with a full width block between stud bays.
6. Position blocking and nailers (for receptacles, switches, cabinets, tubs, cross-walls, etc.) as required and attach using nails.
7. Move second top plates to jig, position for joint overlap and attach using nails.

Drywall Hanging
1. Square frame and clamp to jig to prevent movement.
2. Apply continuous bead of adhesive to frame.
3. Measure and cut drywall and move to jig.
4. Position drywall horizontally with ends on studs, tack using nails, and screw drywall to frame.
5. Sink elevated screws and remove screws that missed studs.
6. Route excess drywall around perimeter of wall and around specified openings.

Figure 3.14
Framing table
used to build
sidewalls

Figure 3.15
Framing table
used to build
sidewalls

Figure 3.16
Framing table
used to build
sidewalls

Figure 3.17
A-frame
framing jig
used to build
sidewalls

3.2.6 Build end walls

Two end walls (short exterior walls) are produced for each module. The length of each wall is equal to the width of the module. The walls are built on a framing jig similar to that used to build partition walls, either a horizontal table (Figure 3.18) or A-frame. A chop saw with an extended length bed is provided to cut top and bottom plates and other components to size. The primary materials and basic processes are similar to those of the sidewall.

Figure 3.18
Framing table used to
build end walls

3.2.7 Build marriage wall

One marriage wall (long wall along marriage line) is produced for each module. The length of the marriage wall is equal to the length of the module. For open floor plans featuring large openings along the marriage line, multiple shorter marriage walls may be used. The marriage wall is built on a long framing jig similar to that used to build the side wall. A chop saw with an extended length bed is provided to cut top and bottom plates and other components to size. The primary materials used include framing lumber (2"x3" or 2"x4"), both pre-cut studs and full length lumber for plates, and drywall (1/2"). The basic processes are similar to those of the sidewall.

3.2.8 Set partition walls

After partition walls are built, they are set on the floor along with tub and shower enclosures (Figures 3.19–3.20). Primary materials used include partition wall assemblies and tub and shower enclosures. The basic process includes:
1. Inspect floor deck for protruding nails and hammer flush as required.
2. Cover floor with plastic to protect from drywall mud droppings.
3. Mark floor showing wall intersections.
4. Position tubs and shower enclosures near point of use.
5. Set walls and attach to floor using nails.
6. Plumb and square walls. Attach intersecting walls using nails along height and gang nails on top.
7. Install tubs and shower enclosures with blocking as required.
8. Cover vinyl flooring with cardboard for protection.

Figure 3.19
Setting partition
walls on module

Figure 3.20
Setting partition walls on
module

3.2.9 Set exterior and marriage walls

After partition walls are set and the side wall, end walls and marriage wall are built, they are set on the module (Figure 3.21). Primary materials used include side, end, and marriage wall assemblies. The basic process includes:
1. Caulk floor where each wall will be set.
2. Set side wall, then end walls, and then the marriage wall and attach to floor using nails.
3. Plumb and square walls. Attach intersecting walls using nails along height and gang nails on top.
4. Apply adhesive foam to all side, end and marriage walls: apply drywall compound to the back of each horizontal drywall seam and spray adhesive foam over drywall seam; spray adhesive foam on one side of all studs and on both sides of studs with a vertical drywall seam.

Figure 3.21
Setting side wall
on module

3.2.10 Install rough electric in walls

After walls are set, electrical boxes and wiring are installed in walls (Figures 3.22–3.24). Primary electrical materials used include the panel box, wiring and wall boxes for switches, outlets, and lighting. Also used are wiring and outlet boxes for telephone, television and data. The basic process includes:
1. Review prints and check circuits and electrical codes.
2. Determine proper lengths and gauges of wire.
3. Assemble and install electrical panel.
4. Locate where all boxes are to be placed.
5. If needed and not pre-installed, install box nailers in walls.
6. Install all boxes.
7. Run wiring. Work with plumber to resolve any conflicts.
8. Wiring runs from the walls into the ceiling are coiled awaiting ceiling set.

Figure 3.22
Panel box subassembly with plug
connectors

Figure 3.23
Electrical boxes and wiring in
exterior wall

Figure 3.24
Wiring in exterior wall

3.2.11 Build plumbing subassemblies

Many subassemblies used in rough plumbing are pre-built and tested in a specialized plumbing workshop (Figures 3.25–3.27). The shop is equipped with a chop saw and other cutting equipment and an assembly workbench. Examples include: water heater supply and temp/pressure relief, kitchen supply, tub/shower diverter, vanity supply, washer, and whirlpool supply. Primary materials include copper and PVC pipe and fixtures of various sizes and other plumbing components. The basic process includes cutting pipe, joining components and pressure testing the subassemblies for leaks. When possible, faucets and plumbing subassemblies are pre-installed on tubs, showers, and water heaters before they are installed in the module.

Figure 3.25
Plumbing subassemblies

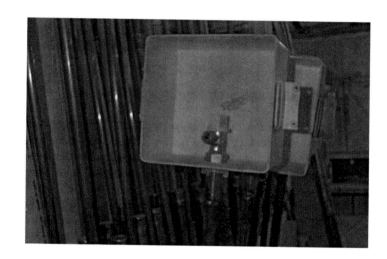

Figure 3.26
Plumbing workshop

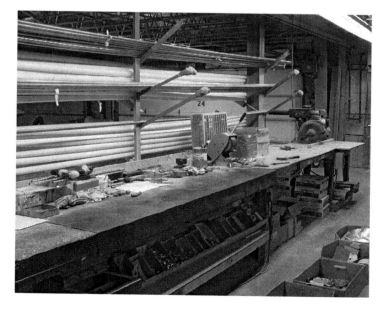

Figure 3.27
Plumbing workshop

3.2.12 Install rough plumbing in walls and tubs

After walls are set, rough plumbing supply and drain lines are installed in walls, tubs and showers (Figures 3.28–3.30). Lines may be used for water, gas, and hot water space heating. Primary materials include copper and PVC pipe and fixtures of various sizes, pre-built plumbing subassemblies and other plumbing components. The basic process includes:

1. Review prints and mark location of drains and supply lines on the floor.
2. Drill holes so that drains and supply lines can be stubbed through floor.
3. Install remaining supply lines, diverters and drains needed by bathtub and showers. Some of these lines may be pre-assembled before loading the tub or shower in the module.
4. Install washer/dryer box.
5. Install any other drains and supply lines in walls.
6. Work with electrician to resolve any conflicts.
7. Install safety plates as required where lines are run through walls.

Figure 3.28
Drain line in wall with vent stack

Figure 3.29
Supply and drain lines on tub

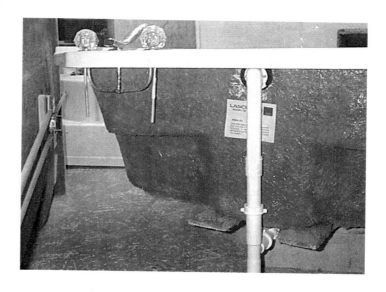

Figure 3.30
Supply and control lines on
shower

3.2.13 Build subassemblies for roof

Several subassemblies are used to build the roof. Trusses provide the structural frame for the roof assembly. Roof trusses used for modular housing fold to minimize shipping height, while allowing steeper roof pitches. Because of their rigid technical requirements driven by the unique home design, the location of the home site and their ability to fold, roof trusses are usually engineered and constructed to order by an external truss supplier. Folding ridge panels are open frame panels used as an extension to the folding trusses for high pitch roofs, allowing the higher pitch without affecting shipping width. Knee walls are open frame panels used to permanently support the folding element of the trusses. Folding eave overhangs are used to form the eave overhangs, allowing generous overhangs without affecting shipping width. Subassemblies for the roof are built in shorter sections and joined as they are attached to the roof. Those subassemblies produced in the factory are assembled on framing jig tables (Figures 3.31–3.32). A chop saw and panel saw are provided in the area to cut components to size. Materials used include: framing lumber (1"x6", 2"x6") and roof sheathing (5/8" plywood or OSB), both precut components and full size. The basic process for each subassembly includes:

Ridge panels
1. Review work order and determine roof pitch, roof length, module width, and dimensions of ridge panels.
2. Mark one stud and cut pattern.
3. For each section of panel, measure length of top and bottom plates, cut to size, mark stud locations (24" on center) and position on framing table.
4. Cut remaining studs using pattern.
5. Position studs on table and attach using nails.

Knee walls
1. Review work order and determine roof pitch, roof length, module width, beams in the roof area and dimensions of knee walls.
2. Mark one stud and cut pattern.
3. For each section of knee wall, measure length of top and bottom plates, cut to size, mark stud locations (24" on center) and position on framing table.
4. Cut remaining studs using pattern.
5. Position studs on table and attach using nails.

Eave Overhangs
1. Review work order and prints to determine if subassemblies are required and, if so, the required dimensions.
2. Measure length of top and bottom plates, cut to size, mark block locations (12" on center) and position on framing table.
3. Rough cut blocks to length and then finish cut to final dimensions.
4. Position blocks on table and attach using nails.

Figure 3.31
Framing table used to build subassemblies for roof

Figure 3.32
Framing table used to build subassemblies for roof (foreground) with completed subassemblies (background)

3.2.14 Build roof/ceiling

A single roof or ceiling is produced for each module (the ceiling replaces the roof atop the lower level of a multi-level structure). The dimensions of the ceiling are equal to those of the module footprint. Roof dimensions include gable overhangs and the eave overhang, unless an eave overhang subassembly is used. For simplicity, the roof/ceiling is referred to as simply the roof in the remainder of the discussion. The roof is built on a framing jig table (Figures 3.33–3.35). The table typically has a solid top surface so that drywall will lie flat for assembly. A chop saw with an extended length bed is provided to cut perimeter rails and other components to size. Primary materials used include pre-built trusses, knee walls, framing lumber (2"x6"x14', 2"x8"x14' and 2"x12"x14' full length lumber and pre-cut joists), and drywall (1/2" drywall in 4'x8' and 4'x14' dimensions). The basic process includes:

1. Position drywall, finish side down, on table.
2. Mark drywall for component locations and cut dimensions, including cutouts, and cut drywall to size.
3. Position trusses, joists, rough openings, wall nailers, ledgers and joist hangers on drywall.
4. Position and nail perimeter framing.
5. Square roof.
6. Prepare adhesive foam machine, apply adhesive foam along the base of each truss and other components, and clean adhesive foam machine.
7. Install knee walls.

Figure 3.33
Framing table used to
build roof (note that
crane is used to
replenish drywall to
material bridge)

Figure 3.34
Framing table used to
build roof (note that
crane is used to
distribute trusses)

Figure 3.35
Framing table used
to build roof

3.2.15 Set roof

After the roof is built, it is set on the module and attached (Figures 3.36–3.38). Primary materials used include the roof. The basic process includes:
1. Lift roof.
2. Verify that walls are plumb.
3. Lower roof onto walls, making sure it is flush with the exterior walls.
4. Flush up sidewall to roof and fasten roof beam to top plate at each stud using nails or other fasteners (lags, straps, etc.).
5. Flush up marriage wall to roof and fasten roof beam to top plate at each stud.
6. Flush up end walls to roof and fasten roof beam to top plate at each stud.
7. Plumb and fasten all interior walls.
8. Remove roof from crane.

Figure 3.36
Setting the roof on module

Figure 3.37
Setting the roof
on module

Figure 3.38
Setting the
ceiling on
module

3.2.16 Install rough electric in roof

After the roof is set, electrical boxes and wiring are installed in the roof (Figures 3.39–3.40). Primary materials used include electrical wiring and boxes for lighting, smoke detectors and fans. Also used are wiring for telephone, television and data. The basic process includes:

1. Review prints and check circuits and electrical codes.
2. Determine proper lengths and gauges of wire.
3. Locate where all boxes are to be placed.
4. If needed and not pre-installed, install box nailers in roof.
5. Install all boxes.
6. Run wiring. Work with plumber to resolve any conflicts.
7. Connect ceiling electrical to runs originating in walls.
8. Perform electrical testing before sealing roof and walls, including pre-test, dielectric and polarity.

Figure 3.39
Ceiling with HVAC ducts, plumbing vent stack (foreground), electrical boxes, can lights and wiring

Figure 3.40
Wiring in wall (note
plug connectors
connecting to wiring
in roof)

3.2.17 Install rough plumbing in roof

After the roof is set, rough plumbing vent stacks are installed (Figure 3.39), and the system is tested (Figure 3.41). Primary materials include PVC pipe and other plumbing components. The basic process includes:
1. Review prints and mark location of vent lines.
2. Install vent lines.
3. Tie all vents together (if applicable).
4. Work with electrician to resolve any conflicts.
5. Install safety plate where pipes are run through the walls and roof.
6. Test pipes in walls and roof before sealing.

Figure 3.41
Plumbing drain lines with test apparatus
attached

3.2.18 Insulate roof

After the roof is set and rough plumbing and electric are installed, the roof is sealed and insulated (Figures 3.42–3.43). A roof is insulated according to applicable building codes with upgrades at customer request. A ceiling is insulated with R-19 around the perimeter and R-11 in the center for noise control. Primary materials include fiberglass batt insulation (R-11–R-38). The basic process consists of sealing all penetrations through the roof with caulk and then unrolling and positioning fiberglass batts in joist bays, being careful not to leave gaps or compress the insulation.

Figure 3.42
Installing
insulation in
ceiling

Figure 3.43
Roof with insulation
installed

3.2.19 Sheath and install subassemblies for roof

After the roof is insulated, it is sheathed and, if required, the pre-built eave overhang and ridge panel subassemblies are installed (Figures 3.44–3.46). If the module has a ceiling instead of a roof, a framework to support the ceiling shipping wrap is installed (Figure 3.47). A panel saw is used to cut sheathing to size, and a chop saw is used to cut frame components to size. Primary materials include framing lumber (1"x3" and 2"x4"), OSB or plywood sheathing (5/8"), and pre-built eave overhang and ridge panel subassemblies. The basic process includes:

Sheath roof
1. Install gable end overhangs if required.
2. Snap chalk line for first row of roof sheathing.
3. Adjust roof to be square and plumb on gable ends.
4. Install cross bracing on trusses as required by truss manufacturer.
5. Install cardboard baffles for ventilation as required.
6. Cut sheathing to size as required.
7. Position first row of sheathing and tack it down.
8. Position remaining sheathing.
9. Fasten sheathing using staples as per specified fastening schedule.
10. Install hinge straps as required at folding roof areas.

Install pre-built eave overhangs
1. Fasten hinge straps to sheathing between each truss along eave of roof using staples.
2. Load pre-built eave overhangs onto roof.
3. Starting at one end of the roof, position first section of eave overhang along roof sheathing and flush with end of roof. Verify that section is flush with roof sheathing on top and square with walls.
4. Fasten section to straps using staples.
5. Repeat process until all sections are installed.
6. Install missing sheathing along overhang used for splicing sections.
7. Mark, cut and install end board on the two end sections.

Install pre-built ridge panels
1. Fasten a hinge strap to the ridge end of the folding rafter in each truss. Straps should be installed before sheathing the roof. Fasten using staples.
2. Load pre-built ridge panels onto roof and position along ridge.
3. Starting at one end of the roof, fasten the strap in each rafter to the corresponding stud in the panel section using staples.
4. After all panel sections are attached, splice sections using blocking at butt joints and nailing.

Install frame for ceiling wrap (ceiling only)
1. Cut short components from 1"x3" and 2"x4" lumber, position one of each to form "T" supports for frame, and fasten using staples.
2. Position a "T" support at the center of the ceiling joist every four feet along length of module and fasten using staples.

3. Cut 1"x3" lumber to size, position widthwise across the module at each support stretching from sidewall to marriage wall, and fasten using staples.
4. Cut 1"x3" lumber to size, position so that it runs the full length of the module every 16" along the width of the module, and fasten using staples.

Figure 3.44
Installing sheathing on roof

Figure 3.45
Installing folding eave overhang on roof

Figure 3.46
Installing folding ridge
panel on roof

Figure 3.47
Installing frame for
wrapping ceiling

3.2.20 Shingle roof

After the roof is sheathed, it is shingled (Figures 3.48–3.49). Primary materials include roofing paper and shingles of various styles and colors. The basic process includes:
1. Starting at one end of the roof, position the drip edge along the eave overhang, overlapping pieces. Fasten using staples.
2. If needed, install ice and water barrier up the first three feet on the roof eaves.

3. Cover the rest of the roof with 15# felt roofing paper.
4. Load shingles onto roof and spread out evenly.
5. Install starter strip of shingles upside down (tar strip up) along and overhanging the drip edge for first row sealer. Fasten the strip using staples or roofing nails, depending on sheathing, shingles, and codes.
6. Attach 1"x2" lumber along both gable ends of roof, flush with sheathing.
7. Starting at the starter strip, install shingles one row at a time, overlapping the lower row.
8. Skip rows at hinge points for folding roofs.
9. Cut shingles around any roof openings, pipes, fan vents or roof windows.
10. Trim back shingles on the gable ends using a router.
11. Remove 1"x2" lumber along both gable ends of roof.
12. Position bundles of shiploose shingles on roof.

Figure 3.48
Installing felt paper on roof

Figure 3.49
Installing shingles on roof

3.2.21 Install fascia and soffit

After the roof is shingled, the fascia and soffit are installed (Figures 3.50). Primary materials include fascia and soffit materials. The basic process includes:
1. Position "F" channel along inside of overhang and fasten with nails.
2. Slide cut pieces of soffit into "F" channel and fasten with staples. Do not staple every fifth piece of soffit.
3. Install soffit along entire length of eave.
4. Remove every fifth piece of soffit. This will be shipped loose so that eave can be fastened in place by builder.
5. Load fascia material for length of eave.
6. Starting at one end of eave, measure end box dimensions and bend fascia material to fit box.
7. Place fascia over box, slide into drip edge and fasten using nails.
8. Install remaining fascia along full length of overhang, sliding into drip edge and fastening using nails.

Figure 3.50
Installing soffit on eave of roof

3.2.22 Prep, drop and wrap roof for shipment

After the roof has been shingled, it is prepped for drop, lowered and then wrapped for shipment (Figures 3.51). The ceiling is wrapped after support for the wrap is added. Primary materials include heavy plastic wrap and framing lumber (1"x3"). The basic process includes:
1. Verify that shiploose shingles are distributed along roof.
2. Remove any fasteners protruding through roof that might damage wrap.
3. Position and attach protective material (foam, corrugated, carpet scraps) over sharp edges that might damage wrap and under hinged roof elements as they are folded for shipping.

4. Remove knee walls and other supports and lower roof for shipping.
5. Verify that roof height does not exceed allowable shipping height.
6. Pull wrap for marriage wall, position and tack to roof along length of roof.
7. Pull wrap for roof and drape evenly over edges.
8. Starting at sidewall, roll wrap around 1"x2" strips and fasten to sidewall using nails.
9. Perform the same process on the marriage wall, verifying that wrap is stretched tight.
10. Wrap on the end walls is rolled, stretched and attached in the same way.
11. Cut, fold and tape wrap at corners to provide water tight seal.
12. Run two straps for the length of the roof and secure at end walls.
13. Run straps every 12' across width of roof and attach.
14. Run two straps vertically from top to bottom on front of module and attach.
15. Run tape vertically over edge of roof and down end walls.

Figure 3.51
Wrapping roof for
shipment

3.2.23 Insulate walls

After the rough plumbing and electric tasks are completed in the side and end walls after roof set, the walls are sealed and insulated using fiberglass insulation (Figure 3.52) or insulating foam (Figure 3.53). Walls are insulated according to applicable building codes with upgrades at customer request. Interior bath walls are also insulated before sealing. Primary materials include fiberglass insulation (R-19 and R-21, 16" and 24" batts depending on stud spacing). The basic process for fiberglass insulation includes:

1. Seal all penetrations through the sidewall, end walls and marriage wall with caulk. Include penetrations through top and bottom plates.
2. Glue and screw plywood gussets to the horizontal drywall seams in the sidewall, end walls and marriage wall.

3. Unroll, position and press fiberglass batts in stud bays of the sidewall and end walls. Be careful not to leave gaps or compress the insulation.
4. Some states require that insulation be stapled to studs and insulation joints be taped.

Figure 3.52
Installing batt insulation in wall

Figure 3.53
Spraying foam insulation in wall

3.2.24 Sheath walls

After insulation is installed in the side and end walls, sheathing is installed on the sidewall, end walls and marriage wall (Figure 3.54). A panel saw is used to cut sheathing to size. For the sidewall and end walls, primary materials include plywood or OSB sheathing (1/2"). For the marriage wall, primary materials include plywood, OSB (3/8"), or other sheathing. The basic process includes:

1. Review prints and work order for type(s) of sheathing to use and locations.
2. Snap chalk lines per prints.
3. Figure the width of the sheathing that will be needed and cut to size using panel saw.
4. Apply construction adhesive to studs.
5. Position sheathing, tack in place and then fasten using staples.
6. Route out door and window openings. Recycle as much scrap as possible.
7. Caulk all seams on sheathing and trowel out excess.

Figure 3.54
Sheathing installed on walls with exterior doors staged before installation

3.2.25 Install windows and exterior doors

After walls are sheathed, windows and exterior doors are installed (Figure 3.54). Primary materials include windows and exterior doors of various sizes. The basic process includes:

1. Verify rough opening dimensions.
2. Install air barrier house wrap per manufacturer instructions, if required. Cut out and fasten wrap at each window opening.
3. Install felt paper flashing around openings, if required, and check for any other special requirements (such as for high wind areas).
4. Caulk inside nailing flange of window.

5. Position window inside opening and verify that window is level and plumb. Fasten as required.
6. Seal around windows and doors with air infiltration tape/flashing as required.

3.2.26 Install siding and trim

After windows and exterior doors are installed, siding and trim are installed on the sidewall (Figure 3.55). Siding is not typically factory-installed on end walls and second floor sidewalls, because of potential alignment problems along vertical and horizontal marriage lines. A chop saw is used to cut siding to size. Primary materials include vinyl, wood, and cementitious siding of various styles and colors. The basic process for vinyl siding includes:

1. Install "J" channel trim strips around windows and doors to cover edge of siding.
2. Strike a chalk line up from bottom of floor rim joists to mark first level of siding.
3. Install an edge piece on one side to maintain siding alignment.
4. Position the first section of the first level of siding along the chalk line and butted against the edge piece. Fasten using staples.
5. Complete installation of the first level along the chalk line, overlapping sections as required.
6. Install remaining levels, one level at a time, starting against the edge piece and snapping each section into the locking flange of the installed lower level and fastening. Note that successive levels are offset horizontally.
7. Siding is cut to length around doors, windows and other objects on the exterior of the module. Wiring that penetrates the siding must be pulled through as siding is installed.

Figure 3.55
Installing cedar siding

3.2.27 Hang drywall on walls

After the rough plumbing and electric tasks are completed in the partition walls and after roof set, the walls are sealed by installing drywall, also called back-paneling. Primary materials include both standard and moisture resistant drywall (1/2" drywall, 4'x8', 54"x8', 4'x12', 54"x12', 4'x14', 54"x14'). The basic process includes:

1. Shim around bathtub and shower units as needed.
2. Finish the installation of pocket doors if required.
3. Check wall for square and adjust as needed.
4. Load needed drywall into module – moisture resistant for kitchen/bath and standard for remainder.
5. Cut drywall to fit.
6. Apply construction adhesive to studs and tack up drywall.
7. Fasten drywall using screws.
8. If unit has firewalls or fire ceilings, install additional drywall as required.
9. Cut out plumbing and electrical access as required.
10. Frame and install drywall for kitchen soffits as required.
11. Install corner bead to outside corners, including walls and soffits.
12. Check for damaged drywall and raised screws and make repairs as needed.
13. Clean out module.

3.2.28 Tape and mud drywall

After the drywall is hung, three coats of drywall mud are applied. The first coat is used to embed the mesh tape used to cover all joints (except outside corners where metal corner bead is used). The final coat is a finish coat using a finish drywall mud. The mud applied in each coat must dry before the succeeding coat is applied. Portable heaters and fans are often used to reduce drying time. Primary materials include both standard and finish drywall mud (dry mortar in bags or pre-mix in buckets). The basic process includes:

1. Inspect drywall surface for problems: high screws, adhesive foam penetrating seams, large holes, etc. Repair as required.
2. Mix drywall mud as needed per the manufacturer's instructions.
3. On all joints requiring tape, apply drywall mud, embed tape in mud covering joint, and smooth starting at one end and working the entire length of the joint.
4. Apply first coat of mud over corner bead on all outside corners.
5. Fill all screw holes and other small dents in the surface with drywall mud and smooth.
6. Allow first coat of mud to dry (20 minutes).
7. Apply second coat of drywall mud over first coat and smooth.
8. Allow second coat of drywall mud to dry (20 minutes).
9. Apply third coat of drywall mud (finish coat) over second coat and smooth.
10. Allow third coat of drywall mud to dry (45 minutes).
11. Clean out module.

3.2.29 Sand and paint

After the finish coat of drywall mud has dried, the drywall surfaces are sanded and painted with primer. Spray-painting equipment is used (Figure 3.56). Primary materials include primer paint. The basic process includes:
1. Mask windows and exterior doors.
2. Sand drywall surfaces.
3. Vacuum dust from module.
4. Clean dust from drywall surfaces using wet mop.
5. Inspect drywall surface and repair as necessary.
6. Spray paint drywall surfaces with primer and then roll.
7. Allow paint to dry.
8. Remove masking from windows and doors.

Figure 3.56
Storage tank for
spray paint system

3.2.30 Install cabinets and vanities

After the paint has dried, the cabinets and vanities are installed in the kitchen, laundry room and bathrooms. Primary materials include cabinets pre-built to order by the supplier. The basic process includes:
1. Mark cabinet locations on walls.
2. Verify appliance openings and proper clearances.
3. Mark stud locations on wall.
4. Unpack and inspect cabinets.
5. Hang upper cabinets.
6. Install base cabinets.
7. Install cabinet knobs and handles.

8. Install cabinet shelves and doors.
9. Install toe kicks.

3.2.31 Fabricate and install kitchen countertops

After the kitchen cabinets have been installed, the countertops are fabricated and installed. A workshop equipped with an assembly table, a chop saw and a panel saw is provided for countertop fabrication. Primary materials include framing lumber, plywood or OSB (24"x12', 30"x12', 36"x12'), laminate and countertops pre-built to order by suppliers (for example, Corian). The basic process includes:
1. Verify countertop measurements.
2. If not purchased from supplier, cut components and assemble countertops.
3. Dry fit countertops.
4. Install countertops.
5. Cut out hole for sink.

3.2.32 Build finish plumbing subassemblies

Sinks, vanity tops and hot water heating subassemblies are pre-built in a specialized plumbing workshop (Figures 3.57–3.58). The shop is equipped with a chop saw and other cutting equipment and an assembly workbench. Primary materials include copper and PVC pipe, fixtures of various sizes and other plumbing components. The basic process includes checking the sink and vanity tops for defects, cutting pipe, and installing components. If necessary, separate sinks are installed in the vanity top. If hot water space heating is used, subassemblies are built.

Figure 3.57
Finish plumbing subassemblies staged in plumbing workshop

Figure 3.58
Finish plumbing
subassemblies staged in
plumbing workshop

3.2.33 Install finish plumbing

After the cabinets and vanities are installed and the required subassemblies are pre-built, the sink, vanity tops, toilets and hot water space heating fixtures (if required) are installed. Primary materials include the pre-built sink and vanity top subassemblies, toilets, hot water space heating subassemblies, copper and PVC pipe, fixtures, and other plumbing components. The basic process includes: installing the sink, vanity tops and toilets; connecting to the rough lines already installed; installing any additional lines as required; and installing hot water space heating (if required).

3.2.34 Install finish electric

After the paint has dried, electrical work can be completed. Primary materials include electrical components and fixtures. The basic process includes installing interior/exterior lighting fixtures, switches, outlets (electric, television, telephone, data, etc.), fire alarms, door chimes, strip heating (if required), wall plates, etc. A final electrical test is performed to complete the work.

3.2.35 Build interior door subassemblies

Interior doors are pre-built in a specialized area equipped with a door assembly jig and a chop saw (Figure 3.59). Primary materials include door slabs (different sizes and styles), wood jamb and casing materials, and door hardware. The basic process includes:
1. Install handle set and hinges on the door blank.
2. Attach the door to one jamb using screws.
3. Build the other jamb subassembly complete with doorstop and door strike.

4. Position the jamb subassembly and top piece around the door and fasten using staples.
5. Attach a strip from one jamb to the other across the door to hold it in place.

Figure 3.59
Workstation for building interior door subassemblies

3.2.36 Install interior doors

After the paint has dried and the pre-built doors are completed, the interior doors are installed. Primary materials include the pre-built door subassemblies. The basic process includes positioning the door subassembly, verifying for level and plumb, and fastening using nails.

3.2.37 Install molding

After the paint has dried, interior molding is installed. This includes base molding, crown molding, window trim, chair rail, and cabinet molding. Molding is stocked and cut in a specialized trim shop equipped with a chop saw (Figure 3.60). Primary materials include wood molding of various sizes and styles. The basic process includes:
1. Measure the exact dimension for each piece of molding.
2. Mark and cut molding to size.
3. Position molding and fasten with nails. Only tack base molding so that it can be fastened after flooring is installed.
4. Touch-up molding with caulk.

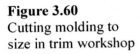

Figure 3.60
Cutting molding to
size in trim workshop

3.2.38 Install miscellaneous finish items

After the paint has dried, finish items such as closet racks, towel bars, and toilet paper holders are installed. Primary materials include closet rack and other miscellaneous finish items in various sizes and styles. The basic process includes measuring closets, marking and cutting rack, and installing the rack and other items.

3.2.39 Install flooring

Vinyl flooring, if required, is installed earlier in the process as the final step in building the floor. After other potentially damaging activities have been completed, other flooring is installed. Flooring options include carpet, wood and ceramic tile. Depending on the needs of the builder, flooring may be installed in the factory or on the construction site. Given the lengthy finish time on the construction site, the potential for damaging factory-installed flooring is high. Even if the builder wants as much flooring installed in the factory as possible, the need for flooring continuity across the marriage line prevents factory installation throughout much of the module. Primary materials include carpet pad and carpet in various sizes and styles, ceramic tile and wood flooring. Carpet is stocked and cut in a specialized carpet area (Figure 3.61). The basic process for carpet installation includes:
1. Install tack strips around perimeter of walls.
2. Install carpet pad and fasten using staples. Tape seams.
3. Cut carpet so that it has excess overhang and position in room.
4. If more than one piece of carpet is required, create seam and attach using seaming tape with seaming iron and roller.
5. Stretch carpet over tacks on tack strip using knee kicker and carpet stretcher.
6. Trim excess carpet.

Figure 3.61
Carpet storage and
cutting area

3.2.40 Load shiploose materials

After finish activities have been completed in the interior of the module, shiploose materials are loaded. Shiploose materials are those materials that will be installed on the home on the construction site by the builder. This includes large subassemblies such as dormers and gable end panels. It also includes exterior materials such as shingles to finish the roof at the hinge points and the ridge, and siding to finish the ends along the marriage line. Interior shiploose materials include drywall and molding to finish the marriage walls, flooring to finish along the marriage line and miscellaneous plumbing, electrical and lighting components. Note that not all shiploose components are loaded inside the module – large subassemblies are loaded on the carrier after the module is loaded, and shingles may be loaded on the roof. Commonly used shiploose items are stocked in a specialized shiploose area. The basic process includes determining shiploose requirements, gathering shiploose materials, loading into the module and securing for shipment.

3.2.41 Factory touch-up

After all potentially damaging activities have been completed inside the module, the module is inspected, necessary repairs completed and the module thoroughly cleaned. The basic process includes visually inspecting the module, repairing any unfinished or damaged areas and thoroughly cleaning the module.

3.2.42 Install plumbing in floor

Immediately before loading the module onto a carrier for shipping, plumbing is completed in the floor (Figures 3.62–3.63). Earlier plumbing tasks left plumbing stubs running through the floor. This task connects these stubs and makes horizontal runs as required. This requires raising the module and placing on portable stanchions. A creeper is then used by the plumber to work underneath. It is also possible to perform this plumbing in the floor immediately after the floor is framed. In this case, the floor is either lifted and placed on longer stanchions (Figure 3.64) or supported over a pit (Figure 3.65). In either case, the floor plumbing is completed with stubs protruding upward through the floor as required. Primary materials include copper and PVC pipe and fixtures of various sizes and other plumbing components. The basic process includes:

1. Install horizontal plumbing lines as required and connect to stubs.
2. Support horizontal runs with metal straps with maximum 4' spacing.
3. Install safety plates as required where lines are run through floor.
4. Seal penetration in floor with caulk.
5. Test all PVC pipe.

Figure 3.62
Plumbing in floor

Figure 3.63
Portable lift for lifting module onto stanchions so that plumbing can be installed in floor (also note creeper used for under-module access)

Figure 3.64
Stanchions used to support floor so that
plumbing can be installed

Figure 3.65
Installing plumbing
and wiring in floor
from a pit under
module

3.2.43 Load module on carrier

After all activities that must be done inside the factory are completed, the module is loaded onto a carrier that is used to transport the module to the construction site (Figure 3.66). Lifting the module so that the carrier can pull underneath for loading is accomplished in several ways: portable electronically synchronized jacks (Figure 3.63), in-floor lifts (Figure 3.67) or heavy duty bridge crane (Figure 3.68). Primary materials include the road-ready carrier and sheathing for covering floor openings. The basic process includes:

1. Get carrier.
2. Lift module so that module clears carrier.
3. Back carrier under module and align per loading requirements such as exterior shiploose materials.
4. Cover any floor openings (such as stairwells) with sheathing from the underside and fasten with staples.
5. Lower module onto carrier.
6. Pull carrier with module out of lift station.
7. Lag module to carrier at floor joists using impact air gun.

Figure 3.66
Carrier used to transport
module to construction site

Figure 3.67
In-floor lift used to lift module
for loading onto carrier

Figure 3.68
Heavy duty
bridge crane
used to lift
module onto
carrier

3.2.44 Final wrap and prep for shipment

Immediately before moving the loaded module out of the factory, the exterior shipping wrap is completed and the module is prepped for shipment (Figure 3.69). Primary materials include 1"x2" lumber to secure wrap on the marriage wall. The basic process includes:

1. Secure the wrap on the marriage wall by rolling it around a 1"x2" on each end of the module and fastening it to the module using nails.
2. Secure the wrap along the bottom of the marriage wall using a similar procedure.
3. If shrink wrap is used, heat after wrapping using heat gun.
4. Secure loose gable sheathing.
5. Check that module meets shipping height requirements and record documentation: serial number, Vehicle Identification Number, Inspection Expiration Date, length of chassis and axle count.
6. Pull module out of factory and stage for shipment.

Figure 3.69
Shrink wrapping module before shipment

3.2.45 Build major shiploose subassemblies

Before shipment to the homebuilder, major shiploose subassemblies are built and loaded onto the carrier (Figures 3.70–3.71). Typical subassemblies include gable end panels, dormers and stairs. Subassemblies are built in a specialized area equipped with framing tables, a chop saw and a panel saw. Materials used include framing lumber, sheathing, shingles, and windows. Basic processes include cutting components to size, framing, sheathing and assembling the subassemblies.

Figure 3.70
Building stairs and dormers in shiploose subassembly area

Figure 3.71
Building gable end
panel in shiploose
subassembly area

3.3 REFERENCES

1. Gianino, A., *The Modular Home*, Storey Publishing, North Adams, MA, 2005.
2. Manufactured Housing Research Alliance, *Develop Innovations in Manufacturing Processes through Lean Production Methods*, U.S. Department of Housing and Urban Development, Affordable Housing Research and Technology Division, Washington, D.C., October, 2005.

CHAPTER 4
DESIGNING THE MODULAR FACTORY

Chapter 4 describes a rigorous structured approach for designing a modular factory. The primary goals include:

- Provide a safe, satisfying and long-term work environment.
- Produce a product that meets customer expectations for variety, customization, performance, delivery time and quality.
- Reduce waste to cut cost, decrease price and increase profitability.

The factory design process, shown in Figure 4.1, is composed of six phases that progress from conceptual to detailed design. The process incorporates key elements of the production strategy developed in Chapter 2, particularly the emerging practices of lean production and mass customization. Although the six phases are shown as serial, the process works best when the phases are performed with some concurrency, or overlap. The process is also iterative – what is learned in latter phases can reshape an initial design decision made in earlier phases.

Figure 4.1
Modular Factory Design
Process

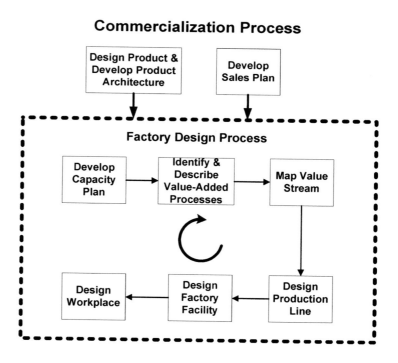

Several caveats should be acknowledged before describing the process further. First, modular producers seldom adopt a rigorous approach to factory design. They commonly cite limited time and resources – the project schedule and budget simply do not permit comprehensive planning. However, an underlying factor is undoubtedly the failure to appreciate the value of more thorough planning, of lean production and of mass customization. This value has been consistently demonstrated in other industries. As a result, modular producers have not been equipped to attain the full promise of modular homebuilding and, consequently, to offer homebuyers compelling value. A final caveat is that a well-planned factory is not necessary nor is it sufficient for business success. Factory design is only one element of many in the overall commercialization process [1]. Success is also highly dependent on the general health of the industry, the local market (homebuyer demand and homebuilding competitors), etc. More than one poorly equipped producer enjoyed financial success in the overheated market of the early twenty-first century, and more than one well-equipped producer has closed in the aftermath. However, a well-equipped producer is more likely to be highly profitable during good times and less likely to be burdened during bad.

4.1 DESIGN PRODUCT AND DEVELOP PRODUCT ARCHITECTURE

Product design defines the product that will be sold over some planning horizon, for example, the next ten years. A product catalog describes standard home designs, discrete options offered, and range of customization allowed. For example, a modular producer will offer a portfolio of standard home designs that represent a range of architectural styles, sizes, levels of finish and cost. The producer will allow homebuyers to customize their homes to meet their unique needs: substituting custom finish elements such as exterior siding, shingles, wood/tile flooring and custom kitchen/bath components; changing room sizes and shapes; adding/deleting rooms; adding a bump-out or a wing; increasing roof pitch; changing from a gable to a hip roof; adding dormers; changing number, location and type of windows; and adding other major design elements such as a porch, garage, or fireplace.

Underlying all other design decisions is the producer's choice of building system. Almost all modular homes are constructed using wood frame construction. However, some producers utilize other building systems including light-gauge steel framing, structural insulated panels (SIPs) and pre-cast concrete panels. Although this discussion assumes wood frame construction, the same basic principles can be applied when using other building systems. The building system coupled with the specific design of a home drives the home's functional performance, influencing structural integrity, energy efficiency, occupant comfort, healthiness of the indoor environment and maintainability. These product design decisions also drive the design and performance of the factory – the choice of production processes, factory configuration and ultimately the cost of building the home. More specifically, the design and performance of the factory are driven by the building system and the size and complexity of the product catalog. In general, for any building system, the greater the size and complexity of the catalog, the more complex and costly it is to build and operate the factory.

The product architecture specifies how the product will be structured – how the home is configured from modules; how a module is configured from major subassemblies; how a major subassembly is configured from minor subassemblies; and how a minor subassembly is configured from building materials and products. Results are often displayed in an indented Bill of Material (BOM), which lists all the building materials, products and subassemblies used to produce each assembly/subassembly. Some features of the product architecture are imposed by the building system. For example, some advanced building systems such as SIPs, light-gauge steel framing and pre-cast concrete panels require that wall panels be used, since they cannot be built in place on the module. Other features of the product architecture are dictated by product design. For example, roof trusses may be required to meet the needs of a complex roof design. Still other features of the product architecture are discretionary – subassemblies can be designed to enhance production performance. For example, most modular producers build the floor, walls and roof as independent subassemblies. If SIPs were to be used, they can be purchased with many value-added features: panels joined to form full wall-length structures, panels cut-to-size and window openings pre-cut and framed.

The wood frame building system utilized by most producers is highly modular (in the product architecture sense). At the elemental level, there are a wide range of compatible materials and products available. They can be mixed-and-matched and configured in a practically limitless number of home designs. They can be interconnected easily, either on the construction site or in the factory. The building system is so modular that product architecture is largely discretionary. Therefore, the designer can create high performance products that the customer wants while the factory can independently structure the product architecture to enhance production performance. Using product architecture to enhance production performance was introduced in Section 2.2.3.1 above and is discussed in greater detail in Section 4.5.2.2 below. It should be emphasized that product architecture is never totally independent of product design, regardless of building system flexibility. Therefore, there should be active collaboration between product design and factory design efforts. Otherwise, potential synergies may be ignored and opportunities lost.

4.2 DEVELOP SALES PLAN

The sales plan documents annual sales projections over the planning horizon. When possible, the sales plan also estimates model mix – the fraction of sales for each standard design, discrete option and level of customization. These projections are then translated into the number and type of modules that must be produced during each year.

4.3 DEVELOP CAPACITY PLAN

The capacity plan is the first phase of the factory design process. Capacity planning was introduced in Section 2.2.1 above. Capacity planning defines how, in the most general sense, capacity will be provided to meet the sales plan – how many hours will be worked at what production rates. The first step is to establish a shift schedule for each year that defines work days per year, shifts per work day and work hours per shift. The average production rate for a year (expressed in modules per hour) is estimated by dividing the number of modules that must be produced (from the sales plan) by the annual work hours (from the shift schedule). The inverse of the production rate is the TAKT time – the average factory cycle time per module (expressed in hours per module). Growth can be accommodated by increasing working hours (adding overtime or shifts) or increasing the production rate (reducing TAKT time). If a phased increase in capacity is planned, then a design year must be chosen to guide initial plant design. The factory design effort must also include a plan for expanding the initial plant design to accommodate any growth of capacity projected over the remainder of the planning horizon.

Several capacity planning issues are particularly relevant to modular producers. Producers have avoided adding a second shift. Their rationale includes:

- Workers on the second and third shifts must be paid a modest shift differential, increasing labor cost per unit.
- It is more difficult to hire and maintain qualified workers for the second or third shift. This may be the result of recruiting from the same labor pool as site construction, where almost all work is performed on the day shift.
- Workers on the second and third shifts do not have the same level of staff support (for example, engineering to resolve design issues).
- It is difficult to manage the transition between shifts – workers must successfully complete the work started by others on the previous shift.

It should be noted that other industries, particularly those that are highly capitalized, routinely operate a second shift. A second issue relevant to modular producers concerns the difficulty in increasing production capacity by reducing the TAKT time. As process cycle time is reduced, the transition (setup) time between cycles becomes a larger percentage of the cycle. Transition activities include raising/lowering catwalks, moving workers and equipment between modules and indexing modules forward on the line. Therefore, a modular producer can expect diminishing returns when reducing the cycle time of a production process, unless transition times are concurrently reduced.

4.4 IDENTIFY AND DESCRIBE VALUE-ADDED PROCESSES

Production processes add value to building materials and products in the BOM, transforming them into completed modules that have been ordered by a customer. The objective of this phase is to identify and describe safe, reliable, cost effective process technologies that can provide the required capacity while accommodating variation. The description of each process includes the building materials, tools/equipment, production methods, labor requirements and cycle time needed to perform the process. The Quality Assurance Manual can serve as a useful starting point for an existing producer. Existing process documentation can be revised to reflect the new product design and architecture, sales plan, capacity plan, and any innovative process technologies being considered.

Alternatives to existing process technologies can be found by investigating process technology suppliers, material suppliers, competitors (and other users), and researchers. These sources can be explored through trade shows, multi-plant industry tours, single plant visits, trade magazines, web searches, equipment directories, material directories, and research reports. Competing process technologies should be compared on the basis of safety, quality, capacity, ability to accommodate variation, and cost. Financial analysis should include both capital costs (equipment, installation, startup) and operating costs (materials, labor, utilities, maintenance).

Elemental processes are consolidated into higher level activities so that they can be more effectively planned and managed. Each higher level activity is performed by a team of workers. Usually the team performs the activity on a single module during a single TAKT cycle. The process of developing the higher level activity structure is called line balancing. It seeks to increase efficiency and reduce overall production cycle time by maximizing the work assigned to each team of workers, while balancing the work content among all teams. This is accomplished by combining processes until some limiting constraint is reached – for example, until the resulting cycle time exceeds TAKT time or working space is insufficient for the workers, equipment and building materials needed. Note that an activity can be limited by other activities that use the same work area at the same time. The processes assigned to a team of workers should be related by common worker skills, process methods, tools/equipment, materials, and precedence relationships. This commonality provides greater focus for the team, encouraging quality and efficiency. It also makes the team more flexible to respond to variation, allowing workers to readily assist team members that are struggling to complete their task within TAKT time. The resulting number of activities usually ranges from 40 to 90, depending on the production rate, the number of workers and the management style of the producer. The production process described in Chapter 3 has 45 activities (Table 3.1).

4.4.1 Estimating labor requirements and cycle time

Labor requirements and cycle time are used throughout the factory design process to assess capacity and financial feasibility. The average work content – the average labor hours required to perform an activity on one module – can be used to estimate both. Assuming that the cycle time needed to perform an activity is inversely proportional to the number of workers performing the activity, these parameters can be estimated as follows:

- Average number of workers required equals average work content divided by TAKT time and rounded up.
- Average cycle time equals average work content divided by the average number of workers required. Note that the resulting average cycle time will be less than or equal to the TAKT time.

For certain activities, these estimates may need to be adjusted to reflect economies (or reverse economies) of scale. For example, a three person crew installing shingles may work disproportionately faster than a two person crew. Processes that have a large fixed time component (often equipment-related) may require disproportionately more workers to reduce cycle time. For example, when using an overhead crane to set the roof, adding personnel beyond a minimum level does not reduce the time required to rig, transport and de-rig the roof.

4.4.1.1 Estimating Work Content

Accurate estimates of average work content are usually not available. It is difficult and expensive to measure the actual labor hours required to perform an activity in the modular factory. Contributing factors include: 1) multiple workers performing the activity, 2) occasional change in worker assignment, 3) frequent movement of workers in and out of modules and between modules, 4) extended activity cycle time, 5) visual obstructions (walls), and 6) frequent delays. These issues are compounded by the large number of activities that are performed in the modular factory. Even if successful in measuring work content, the producer is still challenged by the problem posed by the large number of standard home designs, the unique customization of each home and the resulting variation in work content.

Oglesby et al [2] describe work sampling and time study methods of data collection using direct observation, time-lapse photography and video technologies. Recent research has used time-lapse web-cam technology [3]. A long-term solution may lie in real-time data collection tools like the Status Tracking and Control System prototyped by Mullens [4, 5]. Production workers use wireless laser scanners to report their current work assignments to a host computer, which summarizes the information for each module. Using these tools on a perpetual basis can provide the data needed for real-time shop-floor control and longer-term continuous improvement.

In lieu of actual measurements, average work content can be estimated from current or historical labor levels and TAKT time. The estimate of average work content is equal to the number of workers assigned to an activity multiplied by TAKT time expressed in hours. This estimate assumes that the old and new products and processes are comparable and that TAKT time is representative of average cycle time for the activity. Typical data are summarized in Table 4.1. Estimates are derived from the actual number of workers assigned to activities at five modular producers. These producers have similar products and process technologies. Their production rates range from two to eight modules per day, with low volume and mid-volume firms typically producing higher levels of more elaborate customized products. Each producer has a unique set of higher level activities. However, many activities are common across the firms, and their data can be readily compared. Comparable data for the remaining activities are developed by disaggregating or consolidating activity data to correspond to the activities shown in Table 4.1. The data indicate a wide range in average work content that is closely related to volume. High work content is associated with low volume – all except two of the "high" activity estimates are from producers with a production rate of two modules per day. Most of these "high" estimates are from the same low volume producer, who constructs only a modest level of more elaborate customized products. Therefore, although product mix certainly impacts work content, volume also appears to be a key factor. Therefore, higher volume not only reduces overhead cost, it can reduce labor cost as well. This finding is supported by data from the MHRA benchmarking study (Figure 4.2). The data show a clear trend of work content reduction as volume increases.

Table 4.1 Estimates of Average Work Content by Activity for Five Modular Producers

Act #	Activity	Labor Hr/Module			Notes
		Low	Med	High	
1	Cut to size (Mill)	3	12	17	
2	Build floor	8	8	27	
3	Build window/door opening subassy	2	3	4	
4	Build partition walls	2	5	9	
5	Build side wall	3	5	7	
6	Build end walls	1	2	3	
7	Build marriage wall	2	3	3	
8	Set partition walls	2	3	5	
9	Set exterior & marriage walls	2	4	4	
10	Install rough electric in walls	7	7	17	
11	Build plumbing subassemblies	Insufficient data: Included in Act. 12.			
12	Install rough plumb in wall & tubs	4	6	8	Includes Act. 11,17
13	Build subassemblies for roof	2	4	6	
14	Build roof/ceiling	6	10	13	
15	Set roof	3	4	12	
16	Install rough electric in roof	5	5	7	
17	Install rough plumbing in roof	Insufficient data: Included in Act. 12.			
18	Insulate roof	Insufficient data: Included in Act. 19.			
19	Sheath & install subassy. for roof	5	7	9	
20	Shingle roof	6	7	9	
21	Prep/drop roof & wrap for shipment	2	4	13	
22	Install fascia & soffit	Insufficient data: Included in Act. 24.			
23	Insulate walls	Insufficient data: Included in Act. 24.			
24	Sheath walls	7	14	18	Includes Act. 22, 23
25	Install windows & exterior doors	1	2	20	
26	Install siding & trim	1	4	20	
27	Hang drywall on walls	3	4	8	
28	Tape & mud drywall	14	18	44	
29	Sand & paint	8	12	22	
30	Install cabinets & vanities	4	5	12	
31	Fabricate & install kitchen countertops	2	2	4	
32	Build finish plumbing subassemblies	1	1	2	
33	Install finish plumbing	2	3	4	
34	Install finish electric	3	4	12	
35	Build interior door subassemblies	1	4	6	
36	Install interior doors	1	2	7	
37	Install molding	5	8	24	
38	Install miscellaneous finish items	1	2	4	
39	Install flooring	3	4	4	
40	Load Shiploose	4	5	9	
41	Factory touch-up	9	11	16	
42	Install plumbing in floor	2	3	4	
43	Load module on carrier	0	2	9	Includes Act. 44.
44	Final wrap & prep for shipment	Insufficient data: Included in Act. 43			
45	Build major shiploose subassemblies				No data
	Total		208		

Figure 4.2
Profile of Direct
Labor Hours per
Module for
Modular
Producers [6]

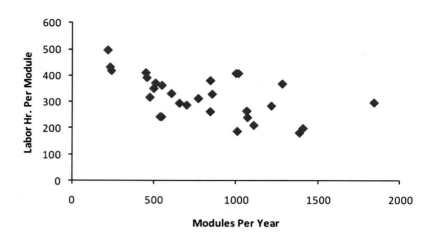

4.4.2 Estimating Cycle Time Variation

Average (static) cycle time is not sufficient to characterize the performance of a candidate process technology in a dynamic production environment subject to large variations in cycle time. For example, work measurement studies of drywall taping at one modular producer documented work content ranging between one and twelve labor hours per module. This variation can disrupt flow and contribute to quality and safety issues. Therefore, a measure of cycle time variation is needed for an objective comparison between competing technologies and effective integration into the production process.

Cycle time variation may be random, emanate from variation in product design (model mix and/or unique customization) or result from wasteful inefficiencies or delays (rework due to poor workmanship or damage, material unavailability or poor quality, design questions or errors, change orders, equipment breakdowns, tool malfunctions). Quantitative data measuring cycle time variation and its root causes are rarely available for modular producers. Measurement is difficult and expensive for the same reasons discussed in Section 4.4.1.1 above. Some insight can be gained by interviewing workers and supervision, particularly for critical tasks.

Product variation is a primary contributor to cycle time variation for many modular production processes. Therefore, a basic understanding of the range of product variation and its potential impact on the production process can provide vital knowledge. Product design elements of interest include the number of different materials required, the number of major parts (windows, interior walls), dimensions (module length/width/height), and design features (kitchens, baths, closets, bumpouts, roof pitch, dormers). Process impacts include additional materials that must be staged, additional tools/equipment needed, additional tasks required, and impact on manpower and cycle time. Some extreme (but common) examples of this variation include:
- A module with no vinyl flooring requires no vinyl installation in the floor framing area.

- A first floor module in a two story home requires a ceiling instead of a roof, which requires no installation of sheathing or shingles.
- A module with no kitchen requires no cabinet installation, while a duplex module may have two kitchens with two sets of cabinets to install.

4.5 MAP THE VALUE STREAM

The value stream map (VSM) is a conceptual model of the production process that organizes the value-added activities in time, establishing when activities occur during the production cycle. It also documents the flow of product and information between activities. If the product remains stationary, it documents the flow of activities to the product. A VSM can be prepared at any level of process detail. A plant-level VSM is used for factory design, indicating flows between major activities in the factory. Value stream mapping plays an important role in the factory design process:

- It promotes communication and a common understanding of the production process among members of the factory design team.
- It allows the design team to see potential waste such as variation, bottlenecks, delays and inventory.
- It enables the design team to identify opportunities to improve the process by restructuring activities, integrating them with other activities and synchronizing the whole to provide reliable, cost efficient capacity while minimizing overall production cycle time.

4.5.1 The VSM

A generic VSM for modular production is shown in Figure 4.3. It can accommodate production of about four modules per eight-hour day. Each quadrant of the VSM is shown in Figures 4.4–4.7. The VSM is based on the activities list shown in Table 4.1. The VSM visually depicts the production (build) process and its environment. The production process is represented on a time scale on the left of the VSM and the environment – the entire enterprise and supply chain – is represented on the right.

4.5.1.1 Production process

The production process is represented on a time scale on the left of the VSM. Each production activity is represented by a rectangular box. An activity is a set of process tasks, usually related, that add value to the product. The activity box also contains key planning data describing the activity including number of workers, number of teams, average cycle time, cycle time variation, rework/defects, delays, etc. Activity data are not shown in Figure 4.3 to conserve space. In practice, the VSM often covers a large "white board" or wall. Activities are organized on the VSM in both the horizontal and vertical dimensions. On the horizontal dimension, activities are organized by module-build and subassembly-build. Module-build activities are organized by the location on the module where the activity is performed: roof, walls (exterior) and interior. On the vertical dimension, activities are organized by their scheduled timing, expressed in TAKT time cycles from the start of module-build. The schedule is constrained by precedence relationships between activities, the limited size of the working area, tool/equipment conflicts and limited access to materials. The schedule also includes intentional idle time for queueing. The overall production cycle time is 23 TAKT cycles, with 18 TAKT cycles needed for module-build.

Most activities are scheduled conventionally – one team of workers performs the activity on one module during one TAKT cycle. Several activities shown in the VSM are not scheduled conventionally: install tubs and rough plumbing in walls, install rough electric in walls, tape and mud drywall, and build subassemblies for roof (one and two). One team of workers performs an unconventional activity on one module over multiple TAKT cycles. The TAKT cycles may be contiguous (install tubs and rough plumbing in walls, install rough electric in walls, tape and mud drywall) or disjointed (build subassemblies for roof – one and two). For a given module, the team performs only part of the activity during each of the designated TAKT cycles. Stated differently, during each production cycle, the team performs part of the activity on multiple modules at progressing stages of completion. Unconventional activities are discussed in greater detail in Section 4.5.2.11 below.

Some activities on the VSM are shown in adjoining boxes; for example, build side walls, build end walls and build marriage walls. This convention is used only to conserve space on the VSM. The activities are unique and independent.

A queue is an intentionally scheduled idle time that interrupts the normal flow of work on the product. A queue is represented by an inventory triangle or a supermarket extended "E". The inventory triangle represents one module or parts for one module that is built-to-order. The supermarket extended "E" represents a supermarket containing one or more parts that are typically common, built-to-stock and pulled as needed. As shown on the VSM, a single module's worth of inventory is planned for every major subassembly and minor subassembly except for pre-cut framing components from the mill. The mill maintains an inventory of common pre-cut framing components in a supermarket and cuts unique custom components as needed. A queue of one module is also planned before roof set. The queue represents a planned delay in the start of roof set and an interruption in the installation of tubs, rough plumbing and rough electric in walls. Queues are discussed in greater detail in Section 4.5.2.9 below.

The physical movement of product between activities is represented by solid flow arrows. A large flow arrow represents module flow, and a small flow arrow represents subassembly flow. The large module-flow arrows shown in the VSM link TAKT cycles instead of specific activities. This indicates that modules flow synchronously at a designated time rather than at the completion of their preceding activities. Precedence relationships between activities are represented by dashed precedence arrows. A precedence arrow indicates that the activity at the tip of the arrow is dependent upon the activity at the tail of the arrow. Precedence relationships are discussed in greater detail in Section 4.5.2.1 below. To simplify the VSM, the small solid flow arrows are also used to represent precedence relationships between their joined activities.

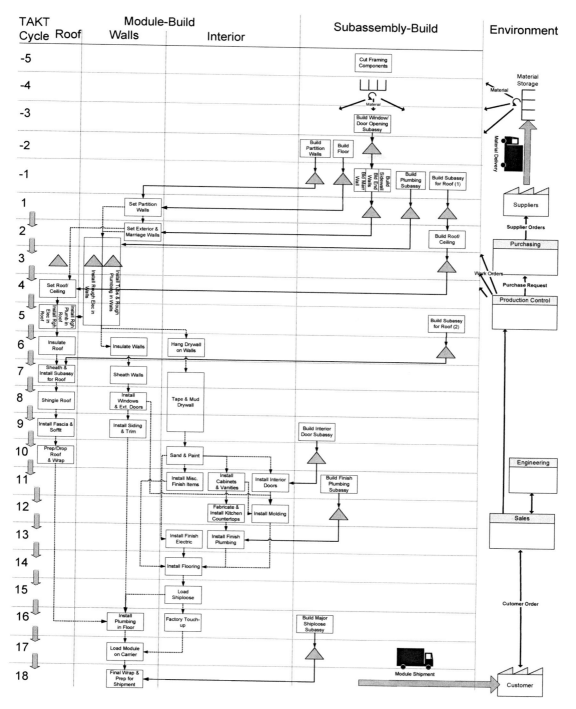

Figure 4.3 Generic VSM for modular production (each quadrant is shown in greater detail in Figures 4.4–4.7)

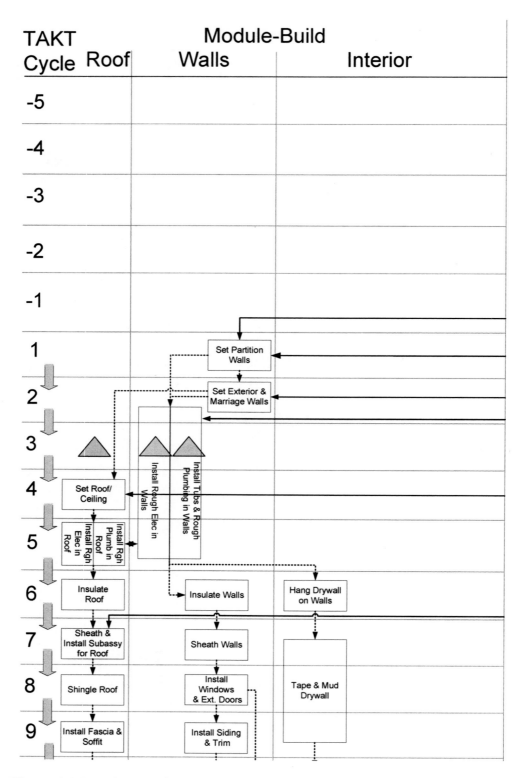

Figure 4.4 Generic VSM for modular production – top left quadrant

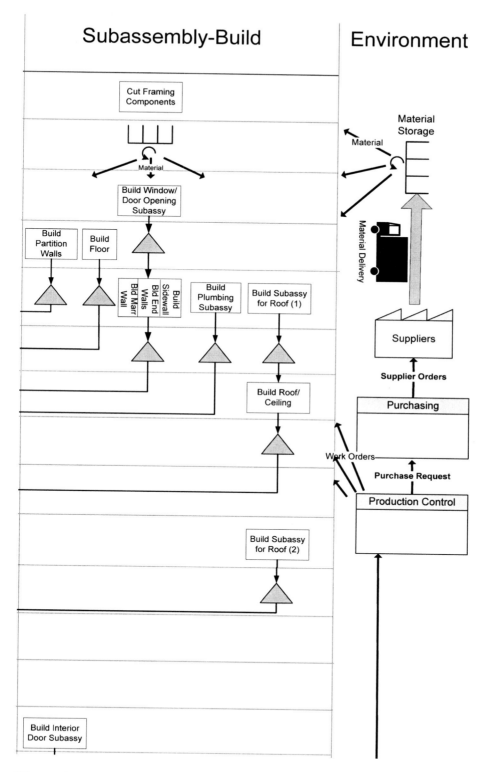

Figure 4.5 Generic VSM for modular production – top right quadrant

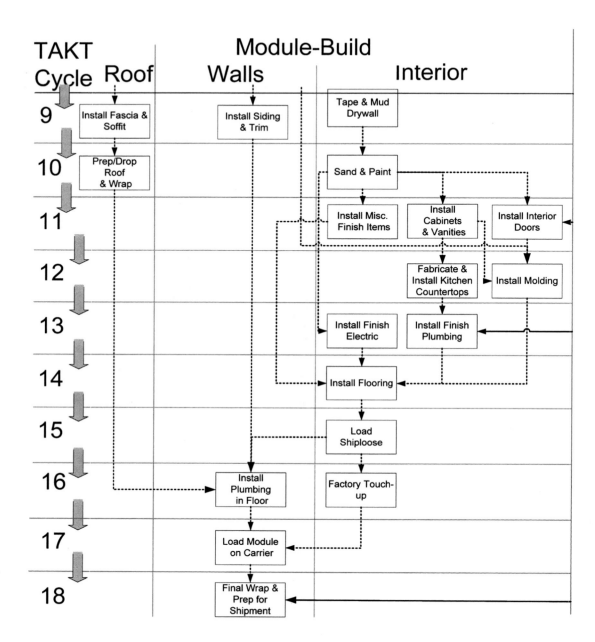

Figure 4.6 Generic VSM for modular production – bottom left quadrant

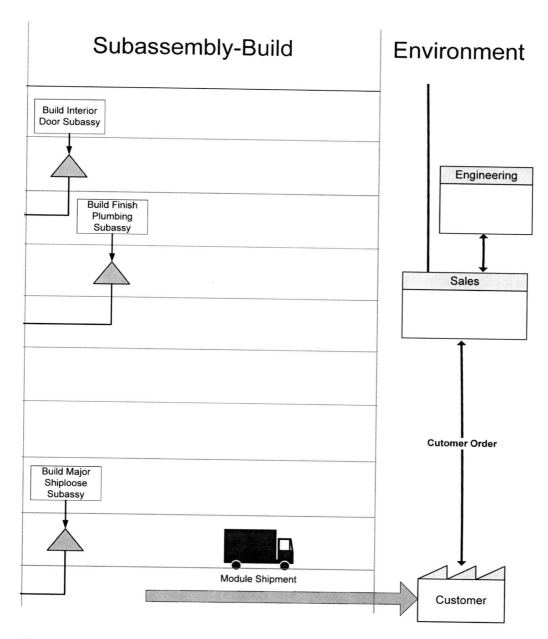

Figure 4.7 Generic VSM for modular production – bottom right quadrant

4.5.1.2 Environment

The right side of the VSM portrays the immediate environment of the production process: the enterprise and supply chain. Activities at this level are not included on the time scale due to their often extended duration. Timing information (cycle time, variation) can be included in their activity boxes.

A customer order flows from the customer to sales. Sales works with both the customer and engineering to fully specify the order. When the order is complete, it is routed to process control. Process control verifies material availability, routes purchase requests to purchasing, schedules the order for production and routes work orders to production. Purchasing places material orders with suppliers, who ship materials to the factory. Materials are inspected upon receipt and placed in inventory. When materials are needed for production, they are transported to production and staged. After completing production, modules are shipped to the customer.

4.5.2 Using the VSM

The VSM is used to structure activities, integrate them with other activities and synchronize them in time. Specific objectives for this and all remaining phases of the factory design process include:
- Provide static capacity – reduce average cycle time for each activity so that it is less than or equal to TAKT time.
- Accommodate variation – reduce cycle time variation. Reduce the impact of any remaining cycle time variation.
- Produce quickly – shorten overall production cycle time.
- Produce efficiently – consume no more resources than needed (floor space, equipment, materials, labor).

The design objectives can be mutually supportive of each other and of higher level goals. For example, providing static capacity and accommodating variation reduce bottlenecks and increase capacity of the overall production process in dynamic day-to-day operations. At the same time they alleviate the "hurry up and wait" syndrome for workers, reducing frustration and exhaustion and enhancing safety, job satisfaction and quality. Producing quickly provides faster response to the customer while increasing efficiency – reducing work-in-process (WIP) inventory and production floor space.

Objectives can also be conflicting and require difficult tradeoffs. For example, many options to provide additional capacity and better accommodate variation require additional resources and extend overall production cycle time. These tradeoffs are at the heart of the factory design effort.

There are many ways to use the VSM to configure the production process so that it better meets objectives. Some options are shown in Table 4.2. Some of these options address activities, exploring their structure, flow and timing. Other options consider product flow. An "X" indicates that the option is likely to support the associated objective. Note that options may support objectives that are not listed. For example, batching an activity allows it to be performed on an off-shift.

Table 4.2 VSM configuration options

Options	Objectives			
	Increase Capacity	Accommodate Variation	Produce Quicker	Increase Efficiency
Expand product architecture	X	X	X	X
Improve process	X	X	X	X
Add resources	X	X		
Replicate activity	X	X		
Decompose activity	X	X		
Synchronize product flow		X		
Develop flexible workforce		X		
Add queue		X		
Develop factories within factory		X		
Consolidate tasks unconventional			X	X
Work in parallel			X	X
Shift activity		X		X
Batch activity				
Build in place				X
Batch line moves				X

These options are described following a brief review of critical path and longest path concepts. While the discussion takes the perspective of designing the factory for increased production capacity and more product variation, the same concepts can also be used to address the reverse scenario – reducing production capacity and product variation.

4.5.2.1 Critical path and longest path

The concepts of critical path and longest path are used repeatedly in the following sections when discussing options for configuring production activities on the VSM. A brief overview of these concepts is useful. For the purposes of this discussion, the duration of each activity and queue is assumed to equal one or more integral TAKT cycles.

The critical path is a construct of a precedence diagram. A precedence diagram is a simplified representation of a process that identifies the activities that comprise the process and their precedence relationships. A precedence relationship is a dependency between two activities that constrains how an activity can progress relative to progress on a preceding activity. Typical precedence relationships include:

- Finish-to-start – an activity cannot be started until the preceding activity is finished. This is the most common form of precedence relationship.
- Finish-to-finish – an activity cannot be finished until the preceding activity is finished.
- Start-to-start – an activity cannot be started until the preceding activity is started.
- Start-to-finish – an activity cannot be finished until the preceding activity is started.

A precedence often results from the product design/architecture. For example, all walls must be set before the roof can be set on the module. A precedence relationship is represented by an arrow from the preceding activity to the affected successor activity. A path is a single strand of connected activities that extends from a "starter" activity (an activity with no precedences) to the final activity on the precedence diagram. There may be many paths sharing many common activities. The duration of a path is the minimum time required to perform all the activities on the path – assuming that each activity starts as soon as its predecessor activity on the path allows. A path is said to be critical if its duration is the greatest on the precedence diagram. More than one path can be critical. The critical path is important since it defines the nominal minimum duration of the overall process, given the activities and their precedence relationships. Activities on the critical path are likewise important, since any delay extends overall process completion. The generic VSM shown in Figures 4.8–4.12 indicates activities on the critical paths. There are many critical paths, sharing many common activities. The duration of these critical paths is 19 TAKT cycles.

The longest path is a construct of the VSM. The VSM represents the actual planned schedule of production activities. It not only recognizes the activities and their precedences identified on the precedence diagram, but also considers other constraints that restrict activity scheduling: the limited size of the working area, tool/equipment conflicts and limited access to materials. The VSM also includes queues between activities. The longest path is the path with the greatest duration on VSM. The longest path is important since it defines the actual minimum time in which a module can be completed – the overall production cycle time. Activities on the longest path are likewise important, since their change can extend overall production cycle time. The generic VSM shown in Figures 4.8–4.12 indicate activities on the longest paths. There are many longest paths, sharing many common activities. The duration of these longest paths is 23 TAKT cycles.

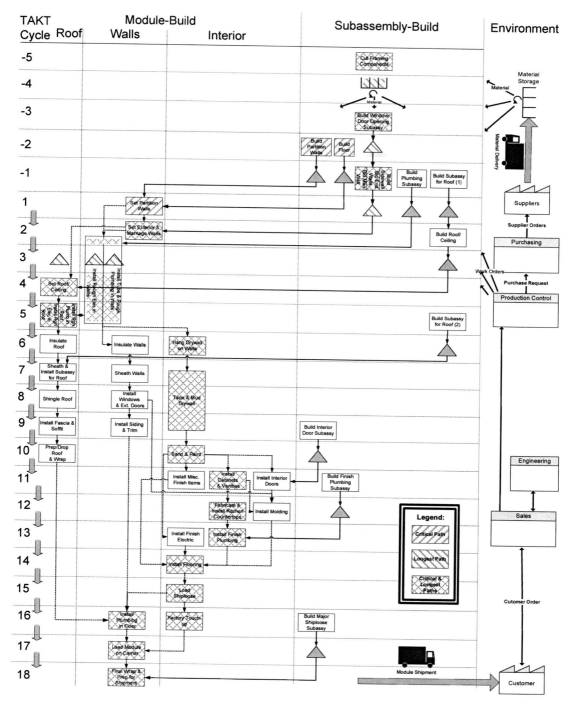

Figure 4.8 VSM showing activities on critical and longest paths

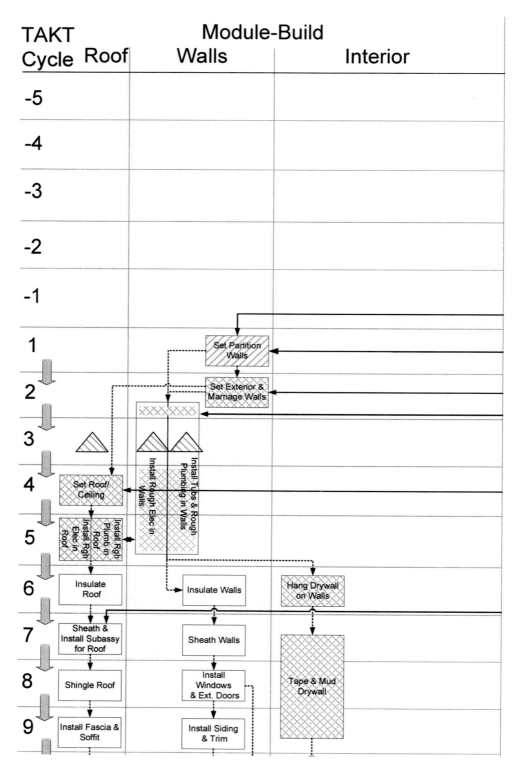

Figure 4.9 VSM showing activities on critical and longest paths: upper left quadrant

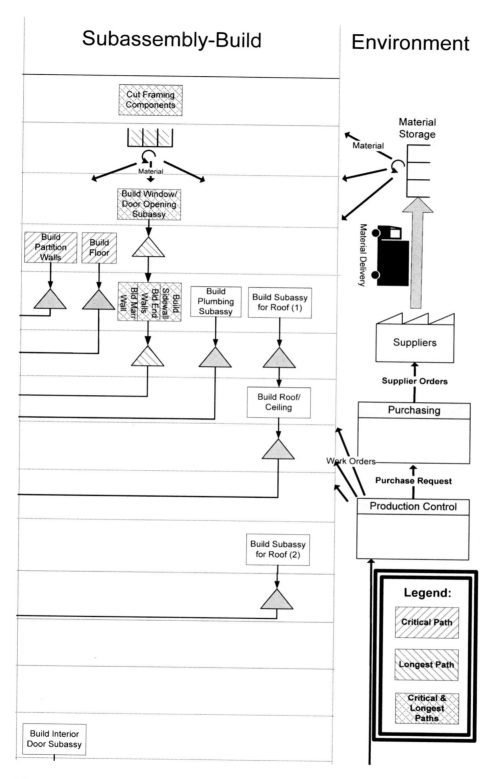

Figure 4.10 VSM showing activities on critical and longest paths: upper right quadrant

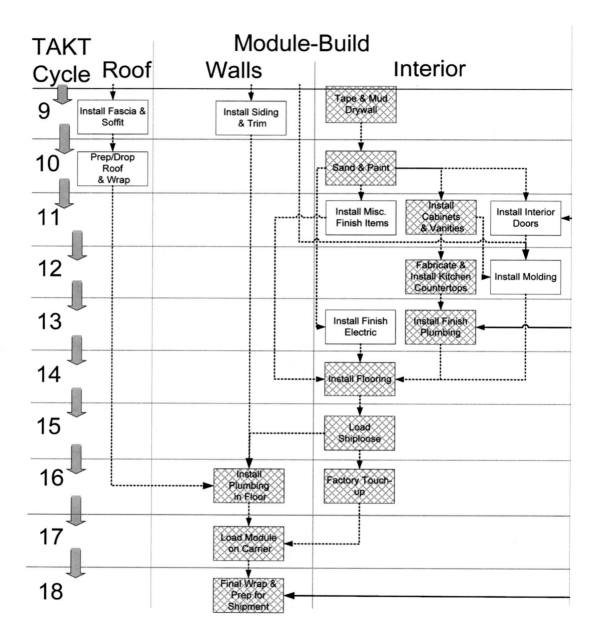

Figure 4.11 VSM showing activities on critical and longest paths: lower left quadrant

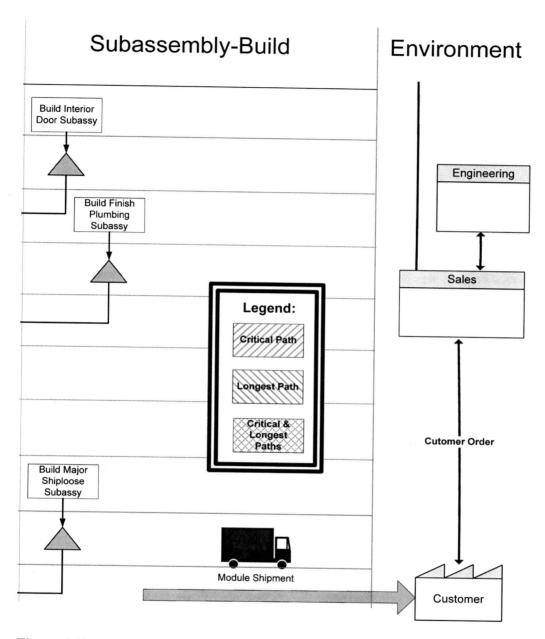

Figure 4.12 VSM showing activities on critical and longest paths: lower right quadrant

It is useful to consider the relationship between the critical path and the longest path. The duration of the longest path is always equal to or greater than the duration of the critical path, since the longest path reflects all the activities and precedences considered in the precedence diagram, plus additional limiting constraints and queues included in the VSM. For example, the duration of the critical paths in Figure 4.8 is 19 TAKT cycles. The corresponding longest paths have a duration of 23 TAKT cycles, including four queues with a duration of one TAKT cycle each. In practice (and as demonstrated in Figure 4.8), critical path activities constitute much of the longest path. This is because: 1) the precedence diagram and VSM recognize the same activities and precedences to determine the critical and longest paths respectively; and 2) the added limiting constraints and queues that are used to determine the longest path on the VSM seldom change it appreciably from the critical path.

While acknowledging the fundamental importance of critical path activities and their close relationship with longest path activities, the remainder of this discussion will focus on activities on the longest path, since they drive overall production cycle time and production resource requirements. Changes to an activity on the longest path can extend the duration of the longest path and, therefore, lengthen overall production cycle time. For example, shifting the activity later on the VSM, extending its duration or decomposing it into multiple activities (each with a duration of one or more TAKT cycles) extends overall production cycle time. In general, overall production cycle time can be reduced by the converse of these measures. However, if there are multiple longest paths, they must all be shortened to effect a reduction in overall production cycle time.

Activities that are not on the longest path are more flexible when configuring the VSM. Their precedence relationships allow them to be shifted on the VSM without directly affecting other activities that are on the longest path and, therefore, without extending overall production cycle time. This flexibility is called "slack". For example, in Figure 4.8 the roof activities after rough plumbing and rough electric can be shifted up to five TAKT cycles later without affecting other activities that are on the longest path. This slack also allows adding a new activity, moving an activity, extending the duration of an activity or decomposing an activity into multiple activities. It is important to note that even though activities may have slack with respect to their precedence constraints, another limiting constraint may still make a change infeasible.

Production resource requirements are also driven by activities on the longest path. To understand the relationship, it is necessary to define the longest path to produce the module and each subassembly. These will be referred to as the longest sub-paths for their respective module/subassemblies. To produce the module or a subassembly, a workstation or queue is needed for each TAKT cycle in the duration of its longest sub-path. Each workstation or queue requires floor space, equipment and WIP inventory. The generic VSM shown in Figures 4.13–4.17 indicate activities on the longest sub-paths. The longest sub-paths for module-build have a duration of 18 TAKT cycles. Therefore, 18 workstations and queues are needed for module-build. The longest sub-path for the production of each subassembly has a duration of two TAKT cycles, one for the build activity and one for a queue. Therefore, one workstation and one queue are needed for each subassembly.

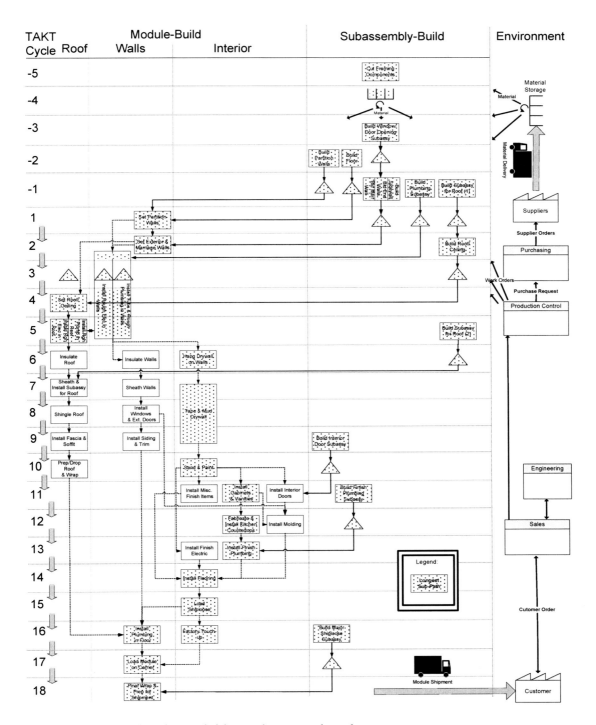

Figure 4.13 VSM showing activities on longest sub-paths

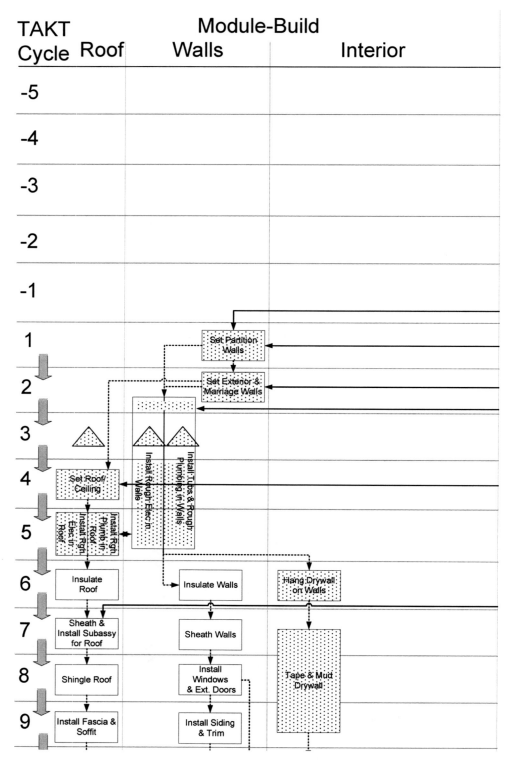

Figure 4.14 VSM showing activities on longest sub-paths: upper left quadrant

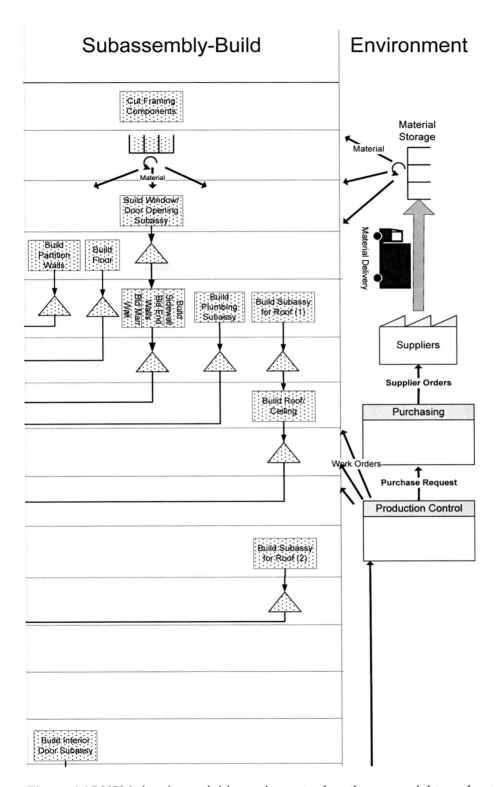

Figure 4.15 VSM showing activities on longest sub-paths: upper right quadrant

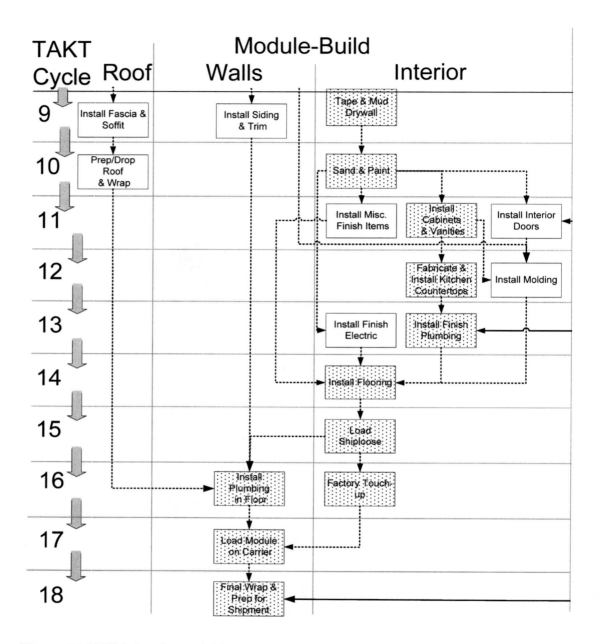

Figure 4.16 VSM showing activities on longest sub-paths: lower left quadrant

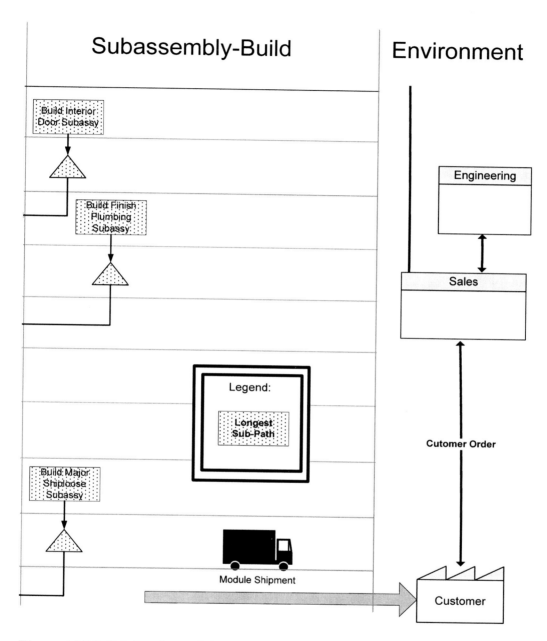

Figure 4.17 VSM showing activities on longest sub-paths: lower right quadrant

Changes to an activity on a longest sub-path can extend its duration and, therefore, require additional workstations/queues to produce the associated module/subassembly. For example, if an activity lies on a longest sub-path, shifting it later on the VSM, extending its duration or decomposing it into multiple activities (each with a duration of one or more TAKT cycles) will require additional workstations/queues to produce the associated module/subassembly. In general, the number of workstations/queues can be reduced by the converse of these measures. However, if there are multiple longest sub-paths for the same module/subassembly, they must all be shortened to effect a reduction in workstations/queues.

An activity that is not on a longest sub-path is performed in parallel with and shares a common TAKT cycle and workstation with another activity that is on the longest sub-path for the associated module/subassembly. The activity also has slack in its schedule, which allows it to be more flexible when configuring the VSM. The activity may be shifted later in the schedule, its duration extended, or it may be decomposed into multiple activities without adding workstations. In addition, new activities may be added or existing activities moved to the sub-path segment without adding workstations. For example, in Figure 4.13 insulation is installed in the roof (from the top) and in the exterior walls (from the exterior) while drywall is being hung on the backside of partition walls (in the interior). Insulation installation can be shifted, extended or decomposed without adding workstations, since the activities are not on the longest sub-path for module-build. It is important to note that even though activities may have slack with respect to their precedence constraints, other limiting constraints may still make a change infeasible.

In day-to-day operations, activities on a longest sub-path are more sensitive to variation. When an activity on a longest sub-path cannot be completed on schedule, the delay is passed directly to downstream activities on the sub-path. If the joining activity at the next higher level of assembly is also on a longest sub-path, the delay is propagated to that sub-path. A queue can be imbedded in a longest sub-path to prevent a delay from affecting downstream activities. Queues are discussed in Section 4.5.2.9 below.

4.5.2.2 Expand product architecture

The product architecture – how the home is configured from modules; how a module is configured from major subassemblies; how a major subassembly is configured from minor subassemblies, etc. – can play an important role in attaining production objectives. Although the product architecture is partially dictated by the choice of building system and product design, there is generally some flexibility to reconfigure the architecture without unduly affecting product design. The wood frame building system utilized by most modular producers allows great flexibility to reconfigure product architecture – to define new subassemblies and reconfigure existing subassemblies – to further production objectives.

A module can be built without subassemblies, while stationary or on a moving line. However, most producers organize assembly into at least two levels: module assembly and major subassemblies (floor, partition walls, side wall, end walls, marriage wall, roof). Each major subassembly is built to its full size, requiring no additional joining when it is set on the module. Because they are full size, most major subassemblies, though similar, are not identical and must be built for the specific order. Minor subassemblies are used sparingly. Most are framing components that feed the major subassemblies or the module itself. These include wall components (window/door openings) and roof components (knee walls, folding ridge panels, folding eave overhangs). The longer roof components are often built to a standard length and then joined when they are installed on the roof or directly on the module. Other minor subassemblies are used by some producers. These include various plumbing subassemblies and interior doors. Because of their smaller scale, minor subassemblies have greater commonality and allow more standardization.

The use of major subassemblies provides important advantages. They are produced in parallel with other production activities, reducing overall production cycle time. They are produced on framing jigs that improve process quality and efficiency. Major subassemblies, particularly those that are two-dimensional (floor, walls), are easier to queue than a module at the same stage of production. Thus, their use facilitates queueing to accommodate variation. The primary disadvantage of using major subassemblies is the need to provide separate workstations where they are built. This involves additional floor space, equipment and WIP inventory. The use of major subassemblies has not been a panacea for modular producers, who continue to cite roof and wall framing most often as production bottlenecks.

The use of minor subassemblies provides similar advantages and disadvantages. The sub-assembly of window openings provides an excellent example. Window openings can be built prior to framing the wall and in parallel with other production activities, reducing overall production cycle time. Window openings are much easier to build on a small specialized jig than on a larger wall jig or directly on the module. Window openings are common for a given window size and ceiling height. This allows the use of a more specialized jig, further improving process quality and efficiency. Common window openings also facilitate a build-to-stock (versus build-to-order) production approach, which can better accommodate variation in the number of windows needed for a module. It is also much easier to queue window openings than walls, further improving process ability to accommodate variation. The primary disadvantage of sub-assembling window openings is the need for a separate workstation. The narrow scope of work also makes it more sensitive to variation. For example, a module requiring no windows generates no work for the workstation. However, the commonality and limited scope of work allow window openings to be efficiently built-to-stock, queued and pulled as needed to build walls.

There are many opportunities to modify the product architecture of a modular home. The configuration of existing subassemblies can be challenged to ensure that they contribute to production objectives. New subassemblies can be explored. Some opportunities include:

- Reduce the scale of a subassembly – for example, building shorter sidewall/endwall subassemblies might increase commonality and, therefore, reduce variation, enhance quality and improve productivity. Shorter subassemblies might be joined when the walls are set on the module.

- Change the components of a subassembly – for example, installing exterior sheathing (instead of drywall) on the sidewall/endwall subassemblies might shorten the subassembly process, strengthen the subassembly and minimize drywall damage that must be repaired. It should be noted, however, that the more common method of installing drywall has important advantages. It allows drywall, rough electric and insulation to be installed outside the module, reducing conflicts with other activities that must be performed inside the module at the same time. More importantly, these "outside" activities are not on the longest path after roof set, allowing drywall finishing to begin sooner. This reduces the overall production cycle time and eliminates a workstation on the line.

- Add more value to a subassembly – for example, installing sheathing and shingles on the roof before setting it on the module might be safer and more efficient than installing after the roof is set. There are several disadvantages with this approach. Additional workstations are required to build the roof, since sheathing and shingling are added to the longest sub-path for roof build. The extra cost of these workstations cannot be mitigated by eliminating the corresponding workstations on the line, since sheathing and shingling were not on the longest sub-path for module-build (see Figure 4.13). From an operational standpoint, the revised process is more sensitive to variation. Delays during sheathing and shingling can bottleneck roof build and pass directly to downstream activities (since sheathing and shingling are added to the longest sub-path for roof build). A roof queue before set can help contain any delays. If the delay does reach roof set, it can bottleneck upstream activities on the line and propagate downstream through the remaining activities (since roof set is on the longest sub-path for module-build). In the current configuration (Figure 4.13), any delays in sheathing and shingling are absorbed by available slack time (since sheathing and shingling are not on the longest sub-path for module-build). A final disadvantage is that it is more difficult to set the roof, install rough plumbing and electric and install insulation after sheathing and shingles are installed.

- Create a new subassembly – for example, creating a new minor subassembly activity to mark top/bottom plates and rim joists centralizes a critical task. Therefore, it is likely to increase efficiency, improve accuracy and reduce variation on the framing jigs. The new subassembly process can be performed in parallel with component cutting and, therefore, will not lengthen overall production cycle time. The new subassembly will require a new sub-path on the VSM, necessitating a new workstation and queue. However, the scale of the subassembly is small, requiring minimal floor space for process and queueing.

- Kit parts – creating a new lower level subassembly to kit the parts needed for an activity can be useful when an activity is complicated by a large number of potential parts, only a few of which are chosen on any given order. Eliminating the complexity of staging and accessing many unique parts shortens the duration of the original activity and eliminates an unnecessary source of variation. A specialized kitting process should also increase the quality and efficiency of parts gathering. This can be particularly useful for activities on a longest sub-path that are likely bottlenecks. Kitting can generally be performed in parallel with earlier activities and, therefore, not lengthen overall production cycle time. Kitting creates a new subassembly and a related sub-path on the VSM, requiring a new workstation and queue. However, the scale of the kitting process is small, requiring minimal floor space for the process and queueing.

- Outsource a subassembly to supplier specialists – for example, using structural insulated panels (SIPs) eliminates exterior wall assembly and improves the functional performance of the home. Outsourcing truss production to suppliers that utilize specialized design software and flexible manufacturing equipment allows custom roof trusses to be produced with high quality and efficiency.

On the VSM, subassembly-build activities are shown on the right side of the production process. On the VSM shown in Figure 4.3, each subassembly-build activity is associated with a unique subassembly. However, a subassembly may be produced using multiple activities performed in series and/or in parallel. To aid visualization on the VSM, activities that are needed to produce a specific subassembly can be outlined or arranged as a unique column. Finally, note that floor build is shown as a subassembly-build activity in Figure 4.3. However, it could also be treated as the first module-build activity.

4.5.2.3 Improve the process

Most modular producers have at least one "problem" activity. The average cycle time for the activity often approaches or exceeds TAKT time, creating a bottleneck in the production system. The same factors that drive longer cycle times cause greater cycle time variation, further disrupting flow. These factors frequently begin with the basic process design and the configuration of the working area. Prior activities may contribute to the problem. For example, excessive damage to drywall in upstream processes often complicates drywall finishing efforts. Direct supervision may exacerbate the situation by not focusing the team on meeting TAKT time, resolving issues quickly when they occur and catching up promptly after falling behind. They may also do little to prevent issues from recurring. Resolutions that require additional resources (overtime or workers) and that lengthen overall production cycle time are often used to treat the symptoms (see Sections 4.5.2.4–4.5.2.6 below). However, before these resource intensive options are considered, the producer should attempt to rationalize the existing process by identifying and addressing the root causes of the problem.

Recent efforts in nine industrialized housing plants have demonstrated that the introduction of lean production techniques can yield striking productivity improvements [6]. Lean production was introduced in Section 2.2.2. Nahmens [7] provides a detailed case study of the improvements in two departments in two HUD Code plants. One department was responsible for applying a finish coat of paint on interior doors. Initially, interior doors were painted after they had been hung in the module. The painting process consisted of 12 tasks as shown in Figure 4.18. Many forms of waste were identified including: 1) movement of painters and their supplies between modules; 2) movement of painters and their supplies between doors throughout a module, 3) movement of painters and supplies to the other side of a door; 4) masking of surrounding areas to prevent paint damage; 5) unmasking of surrounding areas after painting; 6) switching from a roller to a brush to paint doors with a raised profile; and 7) rework due to poor paint finish, paint smudges and damage by other workers.

The improved system centralized the painting operation in an enclosed paint booth near the workstation on the line where doors were installed. The new paint booth accommodated 28 doors. Quick-connect clamps were used to speed setup. The doors swiveled so both sides could be painted with minimal effort. Paint rollers and brushes were replaced by a paint sprayer. The new process consisted of seven tasks as shown in Figure 4.18.

The benefits of the improvements are striking. Spray application is faster and more uniform than the roller. Lengthy travel between modules and doors is eliminated. No masking is required. Standardized procedures in the paint booth are easily reinforced. The paint booth also prevents movement of paint and fumes to other areas of the plant. Cycle time is reduced so much that cycle time variation due to product variation is insignificant – the number and profile of doors in a module has little impact. Door painting is removed from the longest path, allowing subsequent finish activities to begin earlier and reducing overall production cycle time. Even if door painting is delayed, doors can be flexibly installed later in the process. The plant spent $2,000 in material and manpower to build the paint booth. Total labor savings is estimated at $31,500 per year. Defects were reduced from 25% to 5% of doors.

More extensive process improvement can incorporate new technologies. Technology upgrades might start with options that add increased functionality to existing production equipment. For example, a floor framing jig table might be fitted with a material bridge, joist positioning jigs, rail-mounted nail guns and/or a screw gun bridge to increase productivity and reduce cycle time. If additional improvement is needed, a flexible manufacturing system (FMS) should be considered. It typically consists of computer numerically controlled (CNC) machines, linked by an automated material handling and storage system, all under the control of an integrated computer control system. An FMS typically has flexibility in four dimensions: volume, manufacturing processes, product mix, and delivery. FMS technology is capital intensive. This restricts its application in the industrialized housing industry, which has limited capital availability. The variability inherent in the wood frame building system also restricts application. Most FMS applications have been for higher volume wall panel suppliers, where high volume and multi-shift production helps to justify the high cost of the technology. Mullens [8] describes the first successful U.S. application producing interior and exterior wall panels. The system includes semi-automated lumber positioning, lumber nailing, sheathing positioning and sheathing screwing, linked by an automated conveyor system, all under the guidance of an integrated computer control system. Modular producers with their lower volumes have not been able to justify the more expensive, higher capacity equipment. FMS suppliers have not focused on the modular homebuilding market due to its small size.

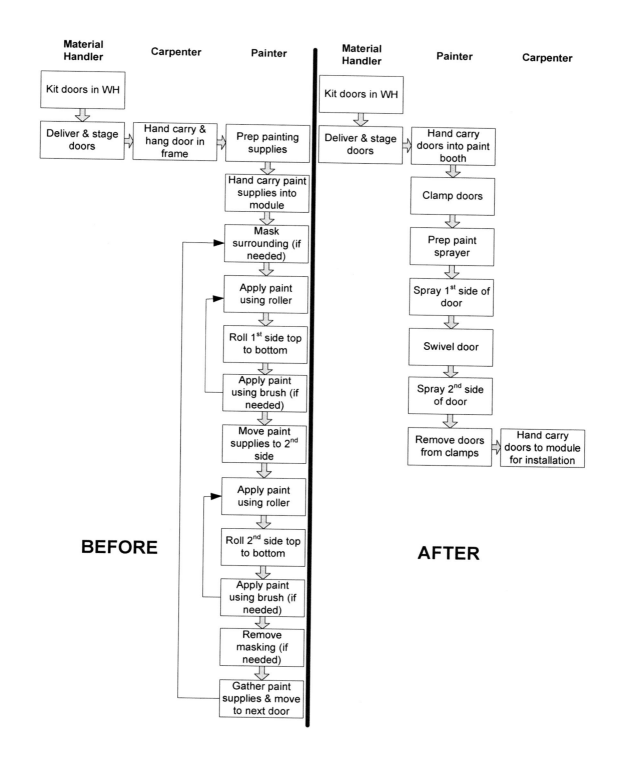

Figure 4.18 Process Flow Charts of Painting Process: Before and After Lean Improvement

4.5.2.4 Add resources

If lower cost approaches to adding capacity or accommodating variation do not yield the desired result, resources can be added. The staffing required to meet a given capacity (to drive activity cycle time below the desired TAKT time) can be estimated from the average work content – the average labor hours required to perform the activity on one module. The estimation of average work content was discussed in Section 4.4.1.1 above. Assuming that cycle time is inversely proportional to the number of workers, these parameters can be estimated as follows:

- Average number of workers required equals average work content divided by TAKT time and rounded up.
- Average cycle time equals average work content divided by the average number of workers required. Note that the resulting average cycle time will be less than or equal to the TAKT time.

For certain activities, these estimates may need to be adjusted to reflect economies (or reverse economies) of scale. For example, a three person crew installing shingles may work disproportionately faster than a two person crew. Processes that have a large fixed time component (often equipment-related) may require disproportionately more workers to reduce cycle time. For example, when using an overhead crane to set the roof, adding personnel beyond a minimum level does not reduce the time required to rig, transport and de-rig the roof.

Average staffing levels may also be adjusted to accommodate variation. This can be particularly useful for key activities on a longest sub-path that experience high cycle time variation. For example, drywall finishing cycle time is a function of wall/ceiling surface area and the number of rooms/closets. Both of these parameters can vary greatly from module to module. To accommodate this variation, maximum expected work content (instead of average work content) might be used to estimate staffing level. This would increase staffing so that cycle time would never exceed TAKT time. The goal is to strategically add capacity to the activity, incurring additional idle time and cost, in order to reduce bottlenecking and delays elsewhere in the process. Recognizing the ability of the overall production process to absorb some variation, one might limit the staffing increase by assuming work content between the average and the maximum.

The primary advantages of adding resources are to increase capacity and better accommodate variation. The primary disadvantage is the associated increase in labor cost.

At some point, it becomes difficult or impossible to drive down cycle time by adding staff. A key limiting factor is the size of the working area: the roof, the interior of a module, a kitchen or bath, or a subassembly framing jig. The area is simply not big enough to support more workers and the materials and equipment they require. Other activities may be performed at the same time and place, putting additional demands on the area. Other limiting factors can include span of supervision, tools/equipment capacity and fixed elements of the cycle time (such as drying time or curing time).

When faced with these limitations, capacity can also be increased by replicating or decomposing the activity. These options are discussed in the following sections.

4.5.2.5 Replicate activity

When the desired capacity cannot be attained by simply adding staff, capacity can be increased by replicating the activity – forming multiple teams that work in parallel. Each of the n parallel teams is staffed by one-n^{th} of the workers. Each team works on every n^{th} module. To maintain capacity, each team must maintain a cycle time of n times the TAKT time. To maintain continuous flow from/to the rest of the production system, one new module enters the replicated activity during each TAKT cycle, and one completed module leaves. Activities are replicated most often to augment capacity in the major subassembly areas: floor, partition walls, marriage wall, end walls, side wall and roof. Industry experience suggests that a second subassembly team (with workstation) is often used when TAKT time is less than two hours. Two to four workers are commonly seen on a subassembly build team, with a maximum of five workers observed. Although seldom observed, module assembly activities can also be replicated. For example, two shingle installation teams might work in parallel, with each team working on every other module and completing the installation in two TAKT cycles. Note that two workstations must be equipped to support this replicated shingle activity.

The primary advantages of replicating an activity are to increase capacity and better accommodate variation. There are several disadvantages. Replicating an activity increases staffing and labor cost. A replicated activity requires one TAKT cycle for each parallel team, increasing the length of all paths and sub-paths on which it lies. If the replicated activity is on a longest sub-path, the number of workstation will be increased. If the activity is on a longest path, overall production cycle time will be extended.

On the VSM, a replicated activity is shown spanning the n TAKT cycles, with the number of teams indicated in the activity box. No replicated activities are shown on the VSM in Figure 4.3.

4.5.2.6 Decompose activity

When the desired capacity cannot be attained by simply adding staff, an activity may also be decomposed into a series of smaller activities. Each resulting activity is assigned to a different TAKT cycle and is performed by its own team of workers. For each resulting activity, cycle time must meet TAKT time, team size must be acceptable, and any other limiting constraints must be satisfied. If the resulting activities are performed on a progressive assembly line, the module/subassembly must be capable of being moved between workstations. For example, instead of replicating the floor build team and workstation to increase capacity, a producer might build the floor progressively in two serial workstations, each staffed with its own team. The floor might be framed at the first workstation and sheathed at the second. When needed, vinyl flooring might also be installed at the second workstation.

The primary advantage of decomposing an activity is to increase capacity and better accommodate variation by reducing the scope of the resulting activities. This reduction in scope also facilitates an increase in focus, which can improve quality and efficiency. Equipment and workers can be specialized. Fewer materials can be staged in larger quantities, closer to their points-of-use. There are disadvantages to decomposition. A decomposed activity requires one TAKT cycle for each resulting smaller activity. In general, this increases the length of all paths and sub-paths on which it lies. If the decomposed activity is on a longest sub-path, the number of workstations will be increased. If the activity is on a longest path, overall production cycle time will be extended. When an activity is decomposed, its closely related progressive work is performed by different teams. This reduces the ownership of each team and may contribute to reduced quality and efficiency. The reduced scope of the smaller activities can also make them more sensitive to variation. In the example above, only sheathing and vinyl flooring are installed at the second floor build workstation. However, many floors do not need vinyl flooring. The result is a smaller, more variable workload that must be performed by a smaller, less flexible team. However, the increased focus might also facilitate process changes to minimize this cycle time variability.

4.5.2.7 Synchronize product flow

Building a product on a progressive assembly line can enhance production performance. Activities are assigned to workstations along the line. Movement of workers is limited to the vicinity of their workstation. A workstation can be specialized to support its assigned activities. Tools/equipment, layout and methods can be rationalized to enhance quality and improve efficiency. Fewer materials are used at each workstation, allowing needed materials to be staged closer to their point-of-use and minimizing staging requirements. Subassemblies can be produced in workstations near their point-of-use on the line, and simple material handling systems can link supplier and user workstations directly.

Cycle time variation can disrupt the flow of product and value-added activity on a progressive assembly line, eroding the potential advantages. To better understand this issue, it is useful to examine how variation affects line performance. For the purpose of this discussion, it is assumed that: 1) each activity is performed in a designated workstation on the line; and 2) product flow between workstations is asynchronous – a product moves as soon as the activities assigned to the workstation are complete and the next workstation is empty. An asynchronous line is often chosen for production systems exhibiting significant cycle time variation, since this assures that an activity is performed in its assigned workstation. When cycle time varies for an activity, the activity is completed before or after the end of its scheduled TAKT cycle. When cycle time exceeds TAKT time, product movement is delayed until the activity is completed. This blocks the line, preventing completed upstream products from moving forward. However, completed downstream products continue to move forward, creating a vacancy at the next workstation. This vacancy is not filled until the delayed activity is completed and the delayed product moves into the next workstation. This also unblocks the line, allowing upstream products to move forward. Note, however, that all activities from the beginning of the line through the vacant workstation are delayed. The original delay propagated immediately to all upstream activities, resulting in idle time and a loss of production capacity.

Consider the downstream impact of the extended cycle time. The vacancy at the next workstation delays the start and subsequent completion of the activities at the workstation. Therefore, a vacancy will be created at the following workstation. The vacancy moves progressively down the line, one workstation for each TAKT cycle. The propagation of the delay downstream results in idle time and loss of production capacity at each successive workstation. When the vacancy reaches the end of the line and the product cannot be completed within its last scheduled TAKT cycle, the result is a loss of production capacity for the entire system. The loss is equal to the excess time required to complete the product. It is possible for the system to recover before the vacancy reaches the end of the line. To recover, all activities at the vacant workstation and upstream workstations must be completed before the end of their scheduled TAKT cycle – before a vacancy occurs at the downstream workstation. Since the start of these activities is delayed, they must all be performed faster. Beginning recovery as soon as possible limits propagation of the vacancy downstream and, therefore, the number of activities involved. If all upstream activities are not successfully completed during the same TAKT cycle, the vacancy reappears downstream of the last activity that is not completed. If the system cannot recover during the production shift, affected activities can "catch up" by working overtime. If the vacancy reaches the end of the line and production capacity is lost, the capacity can be recouped by working the entire system overtime.

When an activity is performed faster than the TAKT time, the team is idled until it begins work on the next product. The team cannot begin this work until the next product is moved into the workstation, which requires that: 1) all activities at the preceding workstation are completed on the next product and 2) the idle workstation is empty – all downstream activities have also been completed early, and downstream products have been moved forward on the line. Since these conditions are highly unlikely, the team is likely to remain idle until the beginning of the next TAKT cycle.

Flow disruptions caused by cycle time variation can be reduced by synchronizing product movement. Synchronous product movement takes place at the same time, at a constant rate. For example, module flow in Figure 4.3 is synchronous. The large module-flow arrows link TAKT cycles (instead of activities), indicating that modules move at the end of each scheduled TAKT cycle, regardless of activity completion. Synchronizing module movement enables the line to continue moving regardless of inevitable variation in activity cycle time. Synchronous line movement requires activities that are mobile – able to move along with products to downstream workstations when cycle time exceeds TAKT time and able to move upstream to the next module when cycle time is less than TAKT time.

Synchronizing product movement provides important advantages. It provides the regular drumbeat needed to synchronize and pace all production activities. This is a cornerstone of lean production, driving process organization and efficiency. Synchronizing product movement also prevents an extended cycle time from disrupting product flow on the line. This, in turn, limits disruption to the flow of activities, minimizing delays, idle time and production loss. When the cycle time for an activity exceeds TAKT time, products continue to flow on the line. This allows upstream activities to continue without disruption. Instead of creating a vacancy at the next workstation, the affected product moves to the next workstation and the affected activity is completed. The time out-of-station for the affected activity is equal to the delay in completing the activity. The out-of-station condition propagates downstream on the line, similar to a vacancy on an asynchronous line and with similar results on idle time and capacity. Recovery is also similar; however, it is much easier. The out-of-station condition starts at the activity responsible for the original delay (rather than at the beginning of the line). Therefore, fewer activities are involved. Recovery can be accomplished incrementally by starting at the activity responsible for the original delay and working downstream. Recovery can be carried out over a period of time, without jeopardizing previously recovered activities.

Synchronizing product movement also allows a team to capitalize when it finishes its activity early (before the end of the current TAKT cycle). The team can move upstream to the preceding workstation on the line and attempt to begin work early on the next product.

There are disadvantages to synchronizing product movement. An activity may not always be performed at a single preferred workstation. Therefore, adjacent workstations must also be equipped to perform the activity. Supervision must actively manage each activity to assure that, when forced out of its preferred workstation, it is able to return within a reasonable period of time. If this does not happen through normal cycle time variation, temporary team augmentation or even overtime may be required. Another consequence of synchronized product movement is that it is possible for a module to be incomplete at the end of the production line, requiring additional work in the yard. This is highly inefficient, erodes quality and is not acceptable on a continuing basis. Again, this requires active management of activities.

It is critical that supervision prevent the production system from stabilizing in an unplanned and sub-optimal condition, where activities are routinely performed out-of-station. Several examples are instructive. In one case, a complex rough electrical system was installed in a flat roof before setting the roof on the module. The installation was performed at a dedicated workstation located on the factory floor, where it could be performed safely and efficiently. Workers had ready access to wiring, fixtures and tools. However, when rough electric delays continued to delay roof set, supervision elected to allow rough electric to be completed after roof set. Rough electric was far more difficult to install after the roof was set on the module, effectively preventing the team from catching up and returning to their dedicated workstation. Management refused to authorize overtime to allow the team to recover. The delay in rough electric also resulted in a delay in completing drywall installation and all downstream finishing activities. As a result, all of these activities were shifted out-of-station and were concentrated in a few workstations at the end of the line. This resulted in severe congestion, poor productivity and rework. In a second case, frequent delays in wall build continued to delay roof set and downstream finishing activities. Delivery schedules were tight, and late deliveries were not considered to be an option. Therefore, there was no time for modules to progress the full length of the line, where activities could be performed at their preferred workstations. Instead, to finish the module by the delivery date, all finish activities shifted upstream on the line, concentrating in a few workstations after roof set. Again, this caused severe congestion, poor productivity and rework.

Several tactics can contribute to activity mobility:
- Establish a zone designating a series of workstations where each activity can be performed. One workstation in the zone may be preferred due to superior access to materials, workshop, or tools/equipment. Note that product cannot move beyond the zone until the activity is complete. If the product is in the last workstation in the zone and the activity is not completed by the end of the TAKT cycle, a vacancy is created downstream and product movement degrades to asynchronous until line recovery. Since asynchronous movement is inefficient and recovery is difficult, it is best to prevent an activity from reaching the end of its zone. Several factors can reduce this risk:
  - Properly size each zone based on the expected variation of the activity. Note that variation in upstream activities can also force an activity out of its preferred workstation.
  - Manage the process to move an activity back to its preferred workstation before it reaches the end of its zone.
- Overlap zones to prevent line length from increasing.
- Equip each workstation in a zone so that the activity can be performed. Some activities, such as installing finish electric, require minimal equipment and materials and can be easily extended to adjacent workstations. Other activities can be more difficult. For example, roof set is often constrained to a single workstation, limited by a single monorail crane that is used to transport the roof to the line. This problem can be resolved by providing a more flexible crane system, capable of serving all workstations in the roof set zone. It may be difficult or impossible to make some activities mobile. For example, the subassembly-build activities shown in Figure 4.3 are not mobile. Each activity must be started and completed in its preferred workstation and move asynchronously. This is indicated on the VSM by flow arrows that link specific activities – a subassembly moves when the previous activity is completed and there is room in the downstream queue.
- Increase flexibility by "relaxing" precedence constraints:
  - Allow workers to begin work in an upstream module before predecessor activities are complete – for example, starting work in a completed room.
  - Allow subassemblies and modules to be moved before an activity is complete. For the roof set example discussed above, if roof build is not complete at the end of the TAKT cycle, allow the incomplete roof to be set and work to be completed after set. If roof set is not complete at the end of the TAKT cycle, allow the module to move forward and set

to be completed after the move. Necessary enablers would include: 1) prioritize tasks required for moving, 2) verify that all tasks required for moving have been completed, and 3) verify that incomplete tasks are completed after the move.

- Provide multiple or mobile locations to stage materials closer to their point-of-use in a zone. For example, a fork attachment for a bridge crane allows shingles to be distributed over any roof located under the crane.
- Provide mobile workshops for smaller, frequently used tools and parts – for example, plumbing and electrical carts allow these activities to be performed efficiently over several workstations.

4.5.2.8 Develop a flexible workforce

Worker mobility can extend beyond performing a single activity at several workstations on the line. A more flexible workforce can help prevent out-of-station disruptions and speed line recovery when disruptions occur. After finishing early, flexible workers might help nearby colleagues that are struggling to complete work on a different activity. Worker mobility can be further enhanced by creating a general category of worker (often termed flex or utility worker) that is responsible for moving throughout the plant and assisting any team needing assistance. Although most modular producers utilize utility workers, they have not been effective in mitigating the effects of variation and preventing bottlenecks. Likely causes include a lack of timely information regarding the need for help and the practical reality that utility workers are usually assigned to cover absenteeism (therefore unavailable to assist other activities that are struggling).

Support systems are needed to develop and utilize a flexible workforce. Cross-training is required to give workers the needed skills for increased mobility. Cross-training can be targeted to link nearby activities that are likely to experience high cycle time variation. Incentives should encourage worker mobility. For example, existing personnel policies often allow workers to leave early (with full pay) after completing their activity on all modules on the daily schedule. This discourages worker mobility. An incentive system based on overall factory performance (quality, cost, production) would better encourage worker mobility. Information systems that forecast likely challenges (extended cycle times) and opportunities (short cycle times) for the upcoming day or week would better prepare supervision and workers. Real time information systems that identify emerging problems (bottlenecks, line vacancies, out-of-station) would help guide immediate action. For example, a red-green light system at each station could be used to signal that a station needs help to complete an activity within the TAKT cycle or when an activity at the station will finish early and can help others. Note that there are many other options for workers that complete their activity early. For example, they might rest or clean/organize the workstation. Guidance should be provided to the worker, both through general guidelines and real time supervision.

4.5.2.9 Add a queue

Queueing can be used to smooth flow on a progressive assembly line. A queue can be added after an activity with high cycle time variability to mitigate the impact on downstream activities. A queue can be added before a bottleneck activity to prevent it from starving and losing vital capacity. Work is not scheduled when the product is in a queue. However, when line movement is synchronous, an incomplete activity can continue after a product is moved into a queue. An activity that is completed early can begin work on the next product in an upstream queue.

A queue can take several forms. A standard queue, represented on the VSM by a triangle, is one product (or parts for one product) that is being built-to-order. A supermarket, represented on the VSM by an extended "E", contains one or more parts that are typically common, built-to-stock and pulled as needed. Queues are controlled to avoid overproduction and the resulting uncontrolled growth in inventory. This is accomplished by linking the upstream supplier activity to downstream user demand. The inventory level for a standard queue in the modular factory is typically controlled by the space and equipment provided to accommodate the large-scale building elements (floor, walls, roof, module). The inventory level for smaller minor subassemblies and other items in a supermarket is set to the minimum required to enable smooth flow. When a specified quantity of an item is withdrawn from the supermarket by a user activity, a visual signal or kanban (card, empty bin, empty cart) is returned to the supplier activity to initiate production of a specified quantity of the item.

A queue can be effective before/after an activity challenged by high cycle time variability and bottlenecking. For example, in Figure 4.3 a queue of one module and a queue of one roof are provided before the crucial roof set activity to buffer it from its highly variable supplier activities. The queue of one module reduces the risk that wall set is not complete by the scheduled start of roof set. Cycle time varies considerably for wall set, particularly for partition walls. Cycle time for building the walls also varies greatly, further increasing the risk that wall set will not be completed on time. The queue of one roof reduces the risk that roof build will not be complete by the scheduled start of roof set. Cycle time for building the roof varies considerably, further increasing the risk that roof set will not be able to start on time. If the module is built on a progressive assembly line, the module queue is typically integrated as a separate station on the line. The roof is difficult to queue because of its shape and the fragility of the drywall on its underside. This requires additional floor space equal to the footprint of the module or suspension above the roof build workstation on stanchions or chains. Two-dimensional subassemblies such as walls are more easily queued, since they can be stacked on edge.

The workload for some activities varies greatly from module to module, but in a more regular pattern. A prime example is cabinet installation. Cabinet installation is performed in modules containing a kitchen – about one-half to one-quarter of all modules, depending on model mix. Assuming that one-half of all modules contain a kitchen and that each kitchen requires cabinet installation, the average work content and the average number of workers required is one-half of that actually needed to install cabinets. Therefore, two TAKT cycles are required to install cabinets in every other module. This delays downstream activities. Downstream delays can be reduced by adding additional workers and reducing cycle time, up to the number of workers required to install cabinets during every TAKT cycle. However, this would result in team idle time for each module not requiring a kitchen. An alternative to this approach is to add a queue after the activity, allowing an average size team to complete the work over several TAKT cycles and to recover when the next module requires no kitchen. This also requires that modules be sequenced for production to spread the design feature.

A queue can also serve as an admittedly inefficient production placeholder for less frequent, labor intensive, custom finish activities, such as installing wood or tile floors. However, there are other options, such as installing on the off-shift, in the yard or on the construction site, which may be better.

There are costs associated with a queue. It adds a TAKT cycle to a longest sub-path and, therefore, adds a queueing station. The queueing station requires floor space, equipment and WIP inventory. If the queue is also on the longest path, it extends overall production cycle time.

4.5.2.10 Establish factories within a factory

A dramatic approach to dealing with product variation is to establish several smaller independent production operations (factories) within the larger factory. Each factory can specialize in a subset of the overall product line: wet (containing kitchen and baths) versus dry modules; small versus large modules; standard versus highly customized/complex modules; fast (low work content) versus slow (high work content) modules; or by specific model families (single family versus condo versus special projects). Each factory runs at its own slower TAKT time based solely on demand for its products. Process activities, workers, layout and equipment are specialized in each factory. Some factories may be configured for maximum flexibility to accommodate larger-scale special projects such as apartments or dormitories. Factories may be completely independent or share common supplier activities such as floor framing, wall framing, roof framing or even complete module framing (with the factories beginning after roof set).

The increased specialization and lower variation in the resulting factories should result in higher quality, greater efficiency and less bottlenecking. However, independent factories are likely to contain redundant resources such as workstations, floor space and equipment. Lower production rates in each factory may lead to lower productivity. Management must be vigilant to counteract this tendency. Specialized factories are also likely to experience greater sales variability. This will require capacity management to meet varying demand in each factory. A flexible workforce, capable of being moved between factories, can help accommodate sales variability.

4.5.2.11 Consolidate tasks unconventionally

Most activities used by modular producers are scheduled conventionally – one team of workers performs the activity on one module during one TAKT cycle. In some cases it is useful to assign tasks to a team, even when these tasks cannot be performed on a module during the same TAKT cycle. Instead, the team performs these tasks on a module over a series of TAKT cycles. For a given module, the team performs only part of the activity during each of the designated TAKT cycles. Stated differently, during each production cycle, the team performs part of the activity on multiple modules at progressing stages of completion.

This unconventional approach can be useful when the producer wishes to use the same team to perform a series of progressive tasks that require common specialized skills, but the tasks cannot be completed on a module in the same TAKT cycle. For example, consider the tape and mud drywall activity. It requires highly specialized skills. The process involves the application of three coats of drywall mud with a drying time of about 20 minutes after each of the first two coats and about 45 minutes after the final coat. A single team cannot efficiently apply all three coats to a module during one TAKT cycle because of the intervening drying times. This becomes increasingly difficult as production rate increases and TAKT time decreases.

A conventional approach for structuring the drywall finishing process would be to decompose the activity into new smaller activities. To maintain efficiency, each coat requires two unique serial activities, application and drying. Each activity requires one TAKT cycle. Therefore, the drywall finishing process requires a total of six TAKT cycles. Since drywall finishing is on the longest sub-path for module-build, six workstations are required. Since it is also on the longest path, drywall finishing constitutes six TAKT cycles of the overall production cycle time.

Using the unconventional approach to drywall finishing shown in Figure 4.3, a single team can apply three coats of mud to three modules located in three different workstations over three TAKT cycles, averaging the required one TAKT cycle per module. This is accomplished by performing a fraction of the work on three different modules during each TAKT cycle: the first coat to module three during its seventh TAKT cycle, the second coat to module two during its eighth TAKT cycle and the final coat to module one during its ninth TAKT cycle. The mud dries on a module while the team is applying mud to the other two modules. Since drywall finishing is on the longest sub-path for module-build, three workstations are required. Since it is also on the longest path, it constitutes three TAKT cycles of the overall production cycle time.

There are many advantages to this unconventional approach. Three fewer workstations are needed. Overall production cycle time is reduced by three TAKT cycles. Since the team works on three different modules during each TAKT cycle, variation in work content per TAKT cycle is reduced. Furthermore, since the larger team is more flexible in its work skills and assignments, it can react more effectively to any variation that does occur. Finally, since the team builds progressively on its previous work, it develops a greater sense of ownership, enhancing quality and improving efficiency. A disadvantage of the unconventional approach is increased movement – the team must move between multiple modules during each TAKT cycle. It is also more difficult to plan and supervise an unconventional activity. The same large team must work on multiple modules at multiple workstations during the same TAKT cycle. As a result, unconventional activities (such as drywall finishing) are often planned with too few workstations and are often a continuing source of delay and disruption in the production system.

The unconventional consolidation approach can also be used to consolidate the production of different subassemblies. For example, in Figure 4.3 the team that builds subassemblies for the roof actually builds two different types of subassemblies during each TAKT cycle:

- For the module in its first TAKT cycle before the start of module-build (TAKT cycle negative one), the team builds subassemblies that will be used during roof build.
- For the module in its fifth TAKT cycle, the team builds subassemblies that will be attached to the roof during the "sheath and install roof subassemblies" activity (after the roof is set).

The advantages are similar to those described above: fewer workstations, less variation in work content and increased flexibility to react to variation.

4.5.2.12 Work in parallel

Performing activities in parallel can shorten overall production cycle time and reduce production resource requirements. The modular production process shown in Figure 4.3 is massively parallel, with many activities occurring at the same time. Opportunities for parallel production can be found across the horizontal dimension of the VSM. The physical size and shape of a module provides many opportunities for parallel processing. A single module can be visualized as three largely independent work areas: exterior, interior and roof. The interior might be further subdivided into wet (kitchen, bath) and dry areas. The interior may be further subdivided by room. Expanding the product architecture by creating a subassembly creates additional opportunities for parallel production, since a subassembly is produced independently of the main assembly.

If two serial activities on a path/sub-path can be performed in parallel, the duration of the path is reduced. If the path is the longest, parallel processing can shorten overall production cycle time. If the sub-path is the longest, parallel processing can reduce the number of workstations required to produce the module/subassembly. Note that if there are other longest paths or sub-paths for the module/subassembly, they must also be shortened to effect the desired improvement. A potential disadvantage of performing multiple activities on the same module/subassembly at the same time is congestion in the work area, which can adversely affect safety, quality and efficiency.

Parallel processing may be limited by precedence relationships between activities, the size of the working area, tool/equipment conflicts, access to materials, etc. It is important to examine the serial activities on longest paths/sub-paths to determine if and under what conditions they could be performed in parallel.

By exploring parallel processing opportunities, it is sometimes possible to decompose an activity (see Section 4.5.2.6) without increasing the number of workstations or extending overall production cycle time. The general strategy is to decompose the activity so that some of the resulting activities can start earlier (rather than later) in the production cycle. For example, consider the installation of rough electric in walls. Although the activity can be started as soon as partition walls are set, it cannot be completed until the remaining walls and the roof are set and wiring has been integrated with wiring in the roof (part of installing rough electric in the roof). A conventional approach might start wall rough electric after the roof is set. However, the scope and scale of wall rough electric is so large, that it will likely need to be replicated or decomposed as capacity is increased and TAKT time decreased. Since wall rough electric is on the longest path and sub-path for module-build, both approaches will, in general, extend overall production cycle time and increase the number of workstations. However, consider decomposing wall rough electric so that some of the resulting activities are started before the roof is set. For example, three smaller wall rough electric activities might be formed: install in partition walls (TAKT cycle two), install in exterior and marriage walls (TAKT cycle four) and complete installation after the roof is set (TAKT cycle five). Note that TAKT cycle three is used for queueing. Each smaller activity has its own team. This approach to decomposition requires no extension of overall production cycle time and no new workstations. This assumes that the resulting activities can be performed without unduly hindering other activities being performed on the module at the same time and without violating any other limiting constraints.

Figure 4.3 shows a similar approach for starting wall rough electric early, but uses an unconventional consolidation approach (see Section 4.5.2.11 above). A single team installs all rough electric in walls. During any given TAKT cycle, the team installs a fraction of the rough electric in each of three different modules: 1) the module in its second TAKT cycle – rough electric installed in partition walls, 2) the module in its fourth TAKT cycle – rough electric installed in exterior and marriage walls and 3) the module in its fifth TAKT cycle – completing rough electric installation after the roof is set. Note that the module is queued during its third TAKT cycle, and no work is planned. The unconventional consolidation approach offers several advantages over decomposition. Since the team works on three different modules during each TAKT cycle, variation in work content per TAKT cycle is reduced. Furthermore, since the larger team is more flexible in its work skills and assignments, it can react more effectively to any variation that does occur. Finally, since the team builds progressively on its previous work, it develops a greater sense of ownership, enhancing quality and improving efficiency. A disadvantage of the unconventional approach is increased movement – the team must move between multiple modules during each TAKT cycle.

4.5.2.13 Shift an activity with slack

An activity that is not on a longest sub-path has added flexibility. Its precedence relationships allow it to be shifted on the VSM without directly affecting activities on the longest path/sub-paths. This flexibility is called "slack". For example, in Figure 4.8 the roof activities after rough plumbing and electric (insulating, sheathing, shingling, installing fascia/soffit, prep/drop/wrap) can be shifted up to five TAKT cycles later, without extending overall production cycle time or adding workstations. An activity may be shifted to: complete the activity as soon as possible; create an idle TAKT cycle to absorb cycle time variation (similar to a queue); balance the activities (workers, materials) assigned to a workstation; or share common equipment, tools and infrastructure (crane, mezzanine) at a workstation.

Slack might also be used to replicate an activity, duplicate an activity, add a new activity or move an existing activity to the path segment with slack. Note that even though an activity may have slack with respect to its precedence constraints, another limiting constraint may make a change infeasible.

4.5.2.14 Batch an activity

It may be useful to perform an activity only during an off-shift. The activity may generate excessive noise/fumes/dust or involve the services of an outside subcontractor with limited availability. To accomplish this, the activity can be batched – delayed during the routine operating shift and performed during an off-shift. A batched activity requires a TAKT cycle for each module to be batched. For example, a factory that operates on one eight-hour shift and has a capacity of four modules per shift (a TAKT time of two hours) accumulates four modules that must be processed in a batch. Therefore, four TAKT cycles are needed. This is far greater than that required for a conventional activity, which requires only one TAKT cycle. The batching penalty is smaller for an activity that requires multiple TAKT cycles (replicated, decomposed, unconventional activities) and for factories with lower production rates. If the batched activity lies on a longest sub-path, an additional workstation is needed for each additional TAKT cycle. The drywall finishing activities (tape and mud, sand and paint) are good candidates for batching. An added benefit of batching the drywall tape and mud activity is that a module is not moved during the critical curing process. This may improve quality and reduce rework. Note that batching several serial activities (such as tape and mud drywall and sand and paint) can further reduce the batching penalty.

4.5.2.15 Build-in-place

Moving a module along a progressive assembly line consumes considerable resources. At the end of each TAKT cycle, workers must exit the module, catwalks must be raised, modules must be moved forward on the line, catwalks must be lowered and workers must enter the next module. As production rate increases (TAKT time decreases), this transition time consumes an even larger percentage of the TAKT cycle. Lean setup reduction techniques (see Section 2.2.2.2 above) can help control transition times. Limiting product movement might also be considered. Several approaches can be used to limit product movement: build-in-place or batching line movement. Build-in-place is discussed in this section and batching line movement is discussed in Section 4.5.2.16 below.

The most drastic way to limit product movement is to build-in-place. The product remains stationary, and all materials and labor flow to the product. Site construction is an example of build-in-place. Modules may also be built-in-place in the factory. Build-in-place may start and end at any point in the process, allowing all modules to utilize common specialty activities. For example, build-in-place might start at the beginning of module-build and use major subassemblies built in common floor framing, wall framing and roof framing workstations. Build-in-place might be delayed until the roof is set, allowing all modules to use a common production line for building and setting major subassemblies. Build-in-place might end prior to loading, allowing all modules to use a common loading station.

Build-in-place does not affect the timing of production activities shown in the VSM or the need for workstations to perform the activities. A build-in-place workstation is needed for each TAKT cycle on the longest sub-path through the build-in-place process. For example, consider the production process shown in Figure 4.3. Assuming a build-in-place approach is used for module-build using the major subassemblies shown, 18 build-in-place workstations are required. After a module is assigned to a build-in-place workstation, all material and teams flow to the workstation according to the schedule shown in the VSM. Note that the large module-flow arrows between these TAKT cycles are removed from the VSM to indicate that the module remains stationary.

The primary advantage of the build-in-place approach is that the product is not moved during production. However, there are disadvantages. All components must be delivered to every workstation. No efficiencies can be gained by locating a supplier workstation near a user workstation, since every workstation is a user workstation. Staging at a workstation is limited. All materials cannot be staged at the same time. Therefore, quantities must be tightly controlled and deliveries closely synchronized with the arrival of teams. Teams must perform their work in every workstation. Team movement must be rationalized by identifying the shortest route (physical distance) through all workstations and consistently flowing teams between adjacent workstations along this route. Specialized facilities and equipment that might be justified to support an activity performed in a unique workstation may not be justified when the activity must be performed in multiple build-in-place workstations. Examples include mezzanines, crane access, a centralized spray paint system and module loading equipment.

A limited build-in-place approach can be used in production scenarios where product movement is constrained. Consider the example described in Section 4.5.2.6 above, where the producer wishes to decompose the floor build activity into two serial activities (framing and sheathing/vinyl flooring) to increase capacity. However, assume that it is difficult to move a framed floor (without sheathing) between the two workstations. Using a limited build-in-place process, two identical workstations could be used, each capable of performing all tasks needed to complete the floor (similar to the replicate approach). However, instead of dedicating a team that builds the entire floor at each workstation, two serial teams could rotate between the workstations. While the framing team works on a floor at one workstation, the sheathing/vinyl flooring team works on another floor at the second workstation. The two teams swap workstations (and floors) each TAKT cycle. Although the floor is built in two separate serial activities, it does not move between workstations. To provide a continuous flow of floors to the line, one new floor order enters the pair of workstations and one completed floor leaves during each TAKT cycle.

4.5.2.16 Batch line movement

A more tempered approach to limiting product movement along a progressive assembly line is to batch the movement. Line movements are reduced by synchronizing and batching movement in multiples of n modules every n^{th} TAKT cycle. For example, modules might be moved forward two workstations every second TAKT cycle. Within the batch cycle, modules remain stationary. Materials and teams flow sequentially to n workstations during the batch cycle. Therefore, an activity that is normally performed in a single workstation is performed in n workstations. A workstation that normally handles activities assigned to one TAKT cycle must handle activities assigned to n TAKT cycles. Batching may begin and end at any point on a progressive assembly line. Batching line movement does not affect the timing of production activities shown in the VSM.

The primary advantage of batching line movement is that it limits product movement along the line. However, it does have disadvantages. It increases the number of workstations required, since the last affected activity requires n workstations to complete its batch cycle. Therefore, n minus one additional workstations are needed to support batch line movement. Other disadvantages are similar to those described in Section 4.5.2.15 above, although on a smaller scale.

Batch line movement is indicated on the VSM by a large module-flow arrow every n^{th} TAKT cycle. The shotgun layout described in Section 4.6.2 offers a related approach to batch line movement.

4.6 DESIGN THE PRODUCTION LINE

The VSM is used to organize activities so that they are synchronized and flow smoothly in time. In this phase of the factory design process, activities are arranged so that product and workers flow smoothly in space. Production line design includes: 1) assigning activities to workstations, 2) arranging the workstations in a layout, and 3) linking the workstations with material handling systems.

Line design begins by determining the number of workstations and queueing positions needed to perform the production activities shown on the VSM. A workstation or queueing position is needed for each TAKT cycle required to produce each module and each sub-assembly. For example, the production system shown in Figure 4.13 requires 18 workstations and queueing positions for module-build. A workstation and a queueing position are needed for each of the 14 subassemblies. The number of workstations and queueing positions is related to production capacity (TAKT time), product design (work content) and product variation (model mix and customization). The relationship between the workstations and queueing positions needed to produce a module and production capacity is shown in Figure 4.19. The modular producers surveyed build modules on a progressive assembly line. Therefore, the workstations indicated include all workstations and queueing positions on their module assembly line. These workstations and queueing positions may lie inside or outside the factory building. The data indicate that, in general, workstations and queueing positions increase with an increase in production capacity. However, the data also show considerable variation. Likely causes of the variation between producers include differing product design, product variation, queueing strategy, production system efficiency, facility size/shape and capital availability.

Figure 4.19
Relationship between module-build workstations/queueing positions and production capacity for modular producers [9]

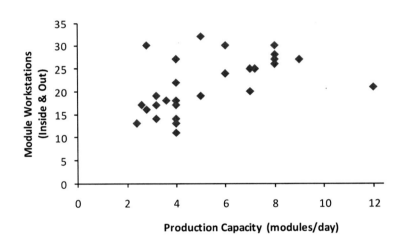

Activities are assigned to workstations based on their timing on the VSM. All activities for a module/subassembly that are scheduled to occur during the same TAKT cycle are assigned to the same workstation. The workstations and queueing positions are then located in space to form a progressive assembly line. This discussion focuses on the arrangement of workstations and queueing positions needed for module-build, since it represents the primary issue in line design. Workstations and queueing positions needed to build major subassemblies (floor, walls, roof) are also considered due to their considerable size and impact on line configuration.

Line design can be classified in several categories including movement, module orientation and general shape. The most basic decision is whether to move the module during the module-build process. The module may be built-in-place or in a series of workstations on a progressive assembly line. The build-in-place approach is discussed in Section 4.6.3 below. Orientation and flow on a progressive assembly line can either be sidesaddle (transverse) or lengthwise. A single lengthwise line is almost never used due to the length of the resulting line. Instead, lengthwise lines are often paired in a "shotgun" configuration, consisting of two shorter parallel lines. Overall line configuration can take the shape of an I, L, T, J, U or almost every conceivable combination. Selected configurations for the line layout are discussed in the following sections. Unless otherwise stated, all configurations accommodate the VSM shown in Fig 4.3. The VSM uses eighteen module-build workstations and queueing positions to accommodate a base production volume of about four modules per eight-hour day. Configurations are also developed for a mid-level production volume of about six modules per day and a high production volume of about eight modules per day. Although VSMs are not developed for the mid-level and high production scenarios, Figure 4.19 can be used to provide a rough estimate of the number of workstations and queueing positions that might be required: twenty-three and twenty-seven respectively. All layouts assume a maximum module length of 65 feet and typical module width of 14 feet.

4.6.1 Sidesaddle line layouts

The sidesaddle line layout is widely used for all production volumes. Modules nearing the end of a sidesaddle production line are shown in Figure 4.20. A typical sidesaddle line layout is shown in Figure 4.21. Sidesaddle flow allows the shortest line length, resulting in a compact factory footprint. The straight flow requires no time consuming turns until a module is loaded onto a carrier. The activity at the last workstation must occur after the module is loaded onto a carrier. Therefore, additional maneuvering with a yard tractor is required, and the location of the last workstation is flexible. Locating it at the end of the line further lengthens the building. Locating it alongside the loading workstation creates a conflict with the resupply of empty carriers to the loading station (Workstation 17). One might also consider streamlining the loading and final activities so that they are consolidated at a single workstation.

Figure 4.20
Modules near the end of a
sidesaddle production line

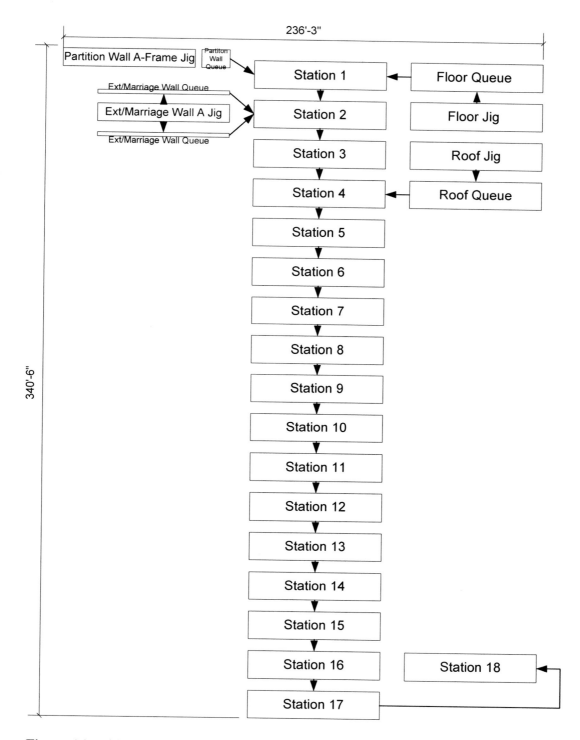

Figure 4.21 Sidesaddle line layout

A straight sidesaddle line allows efficient material flow from primary storage areas to staging locations alongside the line. The line is also easy to expand by simply extending the line in the expansion direction.

The primary weakness of sidesaddle flow is poor access to most parts of the module (except end walls). Movement of workers and materials is long and congested. Access to the module at floor level is through the narrow gaps between modules. Teams install rough plumbing and electric, insulation, sheathing, wrap, windows, doors and siding in these gaps. Large, heavy materials such as drywall, cabinets, countertops, flooring, insulation, sheathing, windows, doors and siding are carried extended distances through the gaps.

Access is worse for activities on the roof: rough plumbing and electric, insulation, sheathing, shingling, installation of fascia and soffit, and prepping roof and wrapping. Access to the roof is typically provided by narrow catwalks. The catwalks are located between modules and extend the entire length of the module. A catwalk can be seen between the first two modules in Figure 4.20. Catwalks are accessed from two mezzanines, one located on each side of the line. The mezzanines extend alongside workstations where access to the roof is required. Most catwalks have limited mobility along the line and must be raised to allow module movement on the line. Catwalks are used as a platform for workers as they perform activities on the roof. Catwalks are also used to move large, heavy building materials used on the roof, such as insulation, roof subassemblies, sheathing and shingles.

The layout illustrates important principles that are used when positioning subassembly workstations and queueing positions. Their orientation is the same as module orientation on the line, negating the need for reorientation. Queueing positions are located alongside their respective supplier workstations and adjacent to their user workstations on the line, simplifying flow. Wall queueing is on edge, rather than flat, improving access and minimizing floor space.

An additional advantage of the subassembly layout shown in Figure 4.21 is that all production activities are performed at floor level, simplifying factory design and product flow. A disadvantage is that there is little room for staging materials (framing lumber, drywall, sheathing, trusses) along the length of the framing jigs. Thus, materials must be staged at the end of the jigs or even out of the workstation, resulting in poor access. Poor material access due to insufficient floor space is a common problem for modular producers. Frequent causes include inadequate initial planning, acquisition of an existing facility that is too small or poorly configured, increased production volume and expanded product line (resulting in more materials that must be stocked). Regardless of the cause, the results are materials that are staged far from their point-of-use, materials that are staged where they are inaccessible (high in a rack or on the bottom of a stack) or materials that are staged in working aisles. Delays accessing these materials reduce production efficiency, add to cycle time variation and contribute to flow disruptions throughout the factory.

Many variations are possible for locating subassembly activities around a sidesaddle line. These include:

- Moving floor build and the floor queue to the front of the line.
- Moving floor build, wall build, or roof build and its queue to the other side of the line or even up onto a mezzanine. It is common to have roof build and its queue located on a mezzanine.
- Moving floor build or roof build to another side of the queue.
- Move the floor or roof queue over the framing jig, supporting it with stanchions or chains.

Line layout and final positioning of subassembly activities should be done in conjunction with detailed design of each workstation (see Section 4.8 below). This includes locating tools and equipment, staging for materials and access aisles.

The sidesaddle layout shown in Figure 4.22 has several additional advantages. The facility requires only two wide (75') bays. A bay is an area between building columns through which product can be moved. Material access is also improved by providing space for material staging on at least one long side of each subassembly framing jig. The layout does, however, extend the length of the facility.

The sidesaddle layout shown in Figure 4.23 provides additional floor space around all sides of the subassembly jigs. The space can be used to stage materials closer to their point-of-use. This can improve material access and relieve congestion, which can be very important as production rate increases and workers are added. The facility requires at least three wide bays. The sidesaddle layout shown in Figure 4.24 provides similar material access and requires a facility with only two wide bays. However, the roof jig is located on a mezzanine, and queues for the floor and roof are located over their jigs.

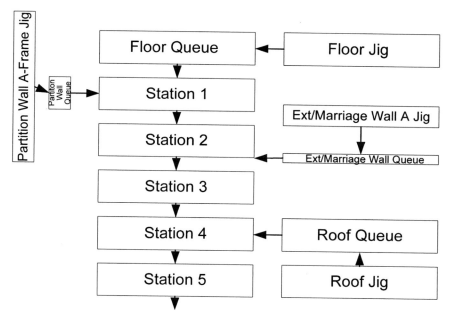

Figure 4.22 Subassembly area for sidesaddle line layout with two wide bays

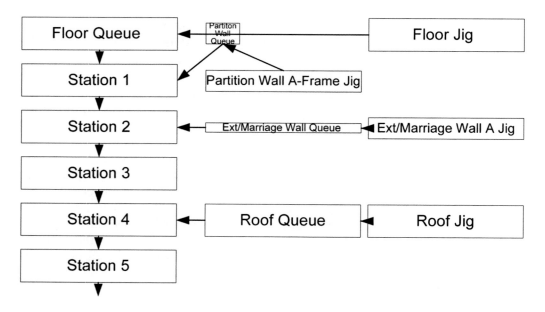

Figure 4.23 Subassembly area for sidesaddle line layout with improved access

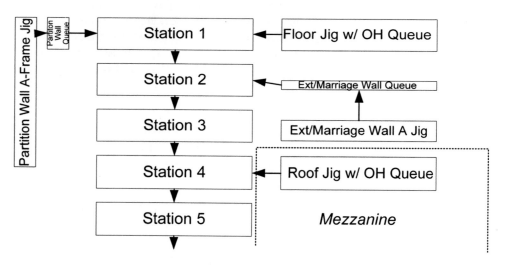

Figure 4.24 Subassembly area for sidesaddle line layout with improved access, two wide bays and mezzanine

Increasing the production rate will require additional workstations, both on the module-build line and for subassembly production. For example, the sidesaddle layout shown in Figure 4.25 might be used for high production volumes. All subassembly production activities are replicated (see Section 4.5.2.5 above). Two workstations are added to the module-build line before roof set. The two workstations might be used for additional queueing after setting walls or to accommodate replicate teams for these activities. Each pair of subassembly workstations flanks the line, resulting in dual queues for all subassemblies except floors, which share a single queue. The resulting layout provides generous floor space around production activities for closer material staging and better access. Floor space might be freed by locating the roof queues directly above the roof build workstations or moving roof build and queueing to a mezzanine.

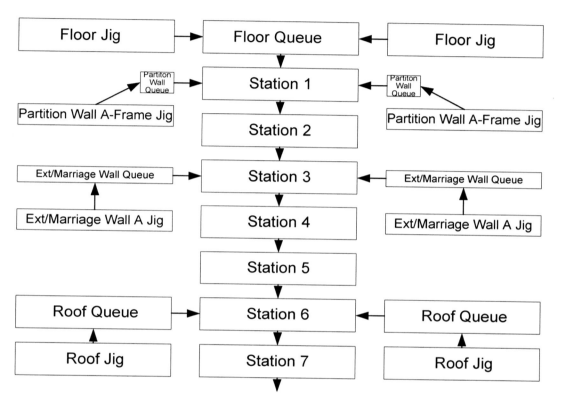

Figure 4.25 Subassembly area for sidesaddle line layout with flanking subassemblies for high production volume

The high volume sidesaddle layout shown in Figure 4.26 consolidates queues to free floor space. It also locates replicated subassembly jigs nearer each other, making it easier for flexible workers to recognize congestion problems on the other jig and move to assist. The sidesaddle layout shown in Figure 4.27 adds production capacity by decomposing the subassembly activities and performing them progressively at two workstations (see Section 4.5.2.6 above). The high volume sidesaddle line shown in Figure 4.28 is arranged in a "J" shape, which allows a more conventional building shape and requires less factory perimeter wall area. The side-by-side workstation configuration also makes it easier for a flexible workforce to recognize congestion problems in another workstation and move to assist. There are disadvantages associated with the J-shaped layout. It entails an internal lengthwise move (from station 17 to station 18), requiring additional time and perhaps additional hardware. If the two parallel lines are adjacent (as shown in Figure 4.28), module access from the common (inner) side is restricted, resulting in less efficient material flow from primary storage areas to staging locations alongside the line. Note that limited floor space for some staging is still available between the lines. Because of the restricted access on the common side, it is useful to leave space on the outside of each line. In Figure 4.28, this is accomplished by ending the line overlap before the roof build area. Note that this could also be alleviated by locating the roof jig on a mezzanine.

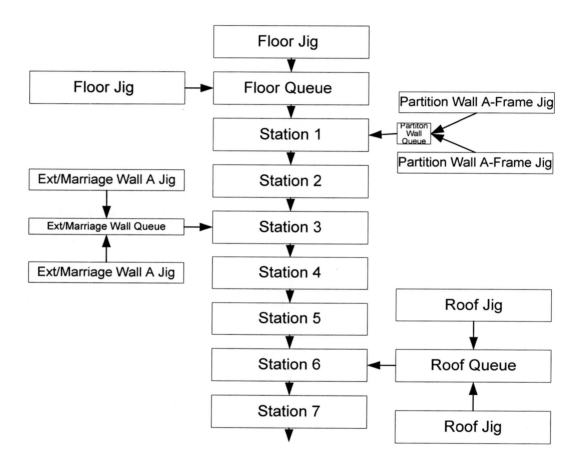

Figure 4.26 Subassembly area for sidesaddle line layout with consolidated queueing for high production volume

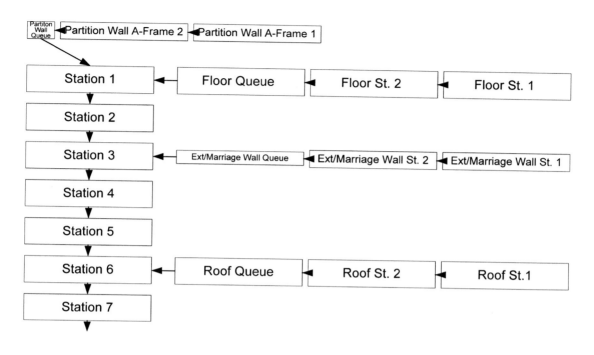

Figure 4.27 Subassembly area for sidesaddle line layout with progressive assembly for high production volume

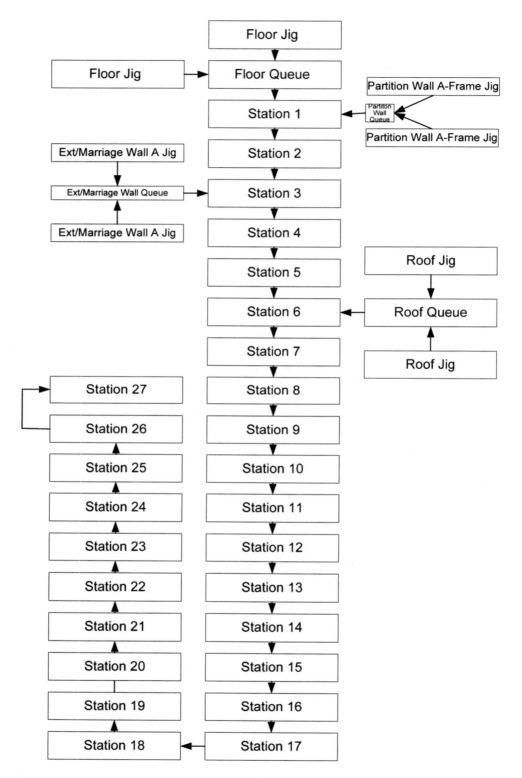

Figure 4.28 J-shaped sidesaddle line layout for high production volume

4.6.2 Shotgun line layouts

A shotgun line layout is sometimes used for lower volume modular production. Modules near the beginning of shotgun production lines are shown in Figure 4.29. A typical shotgun line layout is shown in Figure 4.30. The shotgun line layout uses two straight, parallel lines. Each line contains one-half of the module-build workstations, oriented lengthwise. Modules are assigned to the two lines on an alternating basis – every other module is assigned to a given line. All module-build activities are performed on a module as it flows lengthwise along its assigned line. The cycle time for module movement on each line is twice that of the TAKT time, yielding the required overall production rate. Since a module is built in one-half the number of workstations, twice as much work (twice as many activities) must be performed at each workstation. This is accommodated by the longer dwell time at each workstation – twice the TAKT time (the time required to perform a single activity). For example, the first workstation accommodates partition wall set and exterior/marriage wall set. Partition walls are set during the first TAKT cycle (the first half of the dwell time), and exterior/marriage walls are set during the second TAKT cycle (the second half of the dwell time). Note that the two lines are not independent – they share the same workforce which is driven by the same TAKT time. Each pair of side-by-side complementary workstations is served by the same two sets of complementary teams, who swap workstations at the end of each TAKT cycle.

Figure 4.29
Modules near the beginning of shotgun production lines

Figure 4.30
Shotgun line layout

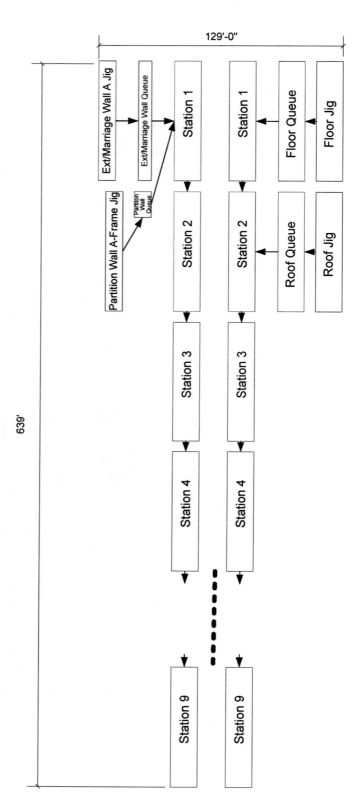

Flow in a shotgun line layout is analogous to batch line movement discussed in Section 4.5.2.16 above, where pairs of modules move together every second TAKT cycle. On a shotgun line layout, the paired stations are located side-by-side on two parallel lines, instead of sequentially on the same line. Module movement on a shotgun line is indicated on the VSM by a large module-flow arrow every second TAKT cycle. To simplify the layout shown in Figure 4.30, the two flows from a subassembly queue to the set workstations (one on each line) are represented by a single flow arrow to the set workstation on the line nearest the queue.

The shotgun layout offers several advantages over a sidesaddle layout. Module movement is reduced by half. Since the workstation dwell time is double that of the sidesaddle line, it is less likely that cycle time variation will force an activity out-of-station. The shotgun orientation with side wall facing outward allows easy access to the side wall. The roof is typically accessed from a mezzanine located on the outside of each line and extending alongside all workstations requiring roof access. No catwalks are needed. The shotgun orientation provides ample space for staging materials alongside the line. As product length varies, it also stabilizes the lineal footage and value of product on the line, preventing a subtle erosion of production capacity. On a sidesaddle line, shorter modules with common framing specifications are sometimes consolidated to reduce this capacity loss.

Like the straight sidesaddle line layout, the straight shotgun layout provides straight line module flow, requiring no time consuming turns. It allows efficient material flow from primary storage areas to staging locations alongside the line. It is also easy to expand.

The straight shotgun layout has a number of disadvantages. It results in a long, narrow building shape that is more difficult to position on a site and expensive to build. This issue becomes more challenging as the production rate increases, workstations are added and lines are lengthened. Access to the interior of the module is poor, limited to entry through a single exterior door on the outside of each line or through the marriage wall between the two lines. Staging between the lines is limited and difficult to resupply from primary storage areas. Fortunately, most interior materials are much smaller than exterior materials (except for drywall, cabinets and countertops) and can be readily staged near the exterior door on the outside of each line. Since all activities are performed on each line, materials staged on the outside of each line are duplicated. This requires additional floor space and possibly materials. Another disadvantage of the shotgun orientation is that the mezzanines needed for roof access are lengthened due to module orientation. Since there are one-half as many workstations on each shotgun line and all activities are performed, each workstation is configured to perform twice as many activities. This is particularly difficult for workstations where the major subassemblies are set, since they are integrated with subassembly workstations and their queues. Increased activity density need not result in increased congestion, since the activities that are performed at the same workstation are performed during two consecutive TAKT cycles.

An advantage of the layout of subassembly activities in Figure 4.30 is that there is ample room for staging materials (framing lumber, drywall, sheathing, trusses) along the length of the framing jigs. Since the shotgun layout is denser with activities along the line, there are fewer good options for locating subassembly activities. These include:

- Moving floor build and the floor queue to the front of the line. This should be considered only under higher production scenarios where a second floor-build workstation is required.
- Moving floor build, wall build, or roof build and its queue to the other side of the line or even up on a mezzanine. It is common to have roof build and its queue located on a mezzanine. At least one producer has located wall build on a mezzanine.
- Moving floor build or roof build to another side of the queue.
- Move the floor or roof queue overhead, supporting it with stanchions or chains.

Increasing the production rate will require additional workstations, both on the module-build line and for subassembly production. For example, the shotgun layout shown in Figure 4.31 might be used for mid-level production volumes. All subassembly production activities are replicated (see Section 4.5.2.5 above). A total of 12 workstations/queueing positions are provided on each line. Exterior/marriage walls are set at a new workstation added to each line before roof set. The new station provides queueing between setting partition walls and setting exterior/marriage walls. This results in the following schedule of activities:

- TAKT cycle one – set partition walls in Workstation 1
- TAKT cycle two – queueing in Workstation 1
- TAKT cycle three – set exterior/marriage walls in Workstation 2
- TAKT cycle four – queueing in Workstation 2
- TAKT cycle five – set roof in Workstation 3

Each pair of subassembly workstations flanks the line, resulting in dual queues for each subassembly. Floor queues are located at the head of the lines. To simplify flow, each subassembly workstation and its queue is dedicated to the nearest line. The layout provides generous floor space around subassembly workstations for closer material staging and better access. Additional floor space might be freed by locating queues directly above the floor and roof build workstations or moving roof build to a mezzanine.

The shotgun line layout shown in Figure 4.32 frees floor space by locating each pair of subassembly jigs on the same side of the line and consolidating their queues. It also locates replicated subassembly jigs nearer each other, making it easier for flexible workers to recognize congestion problems on the other jig and move to assist. A disadvantage of this approach is that subassembly flow is more complicated. A subassembly moving from its queue on one side of the line to the workstations where it is set on the far line must travel over the nearer line.

Figure 4.31
Shotgun line layout with flanking
subassemblies for mid-level
production volume

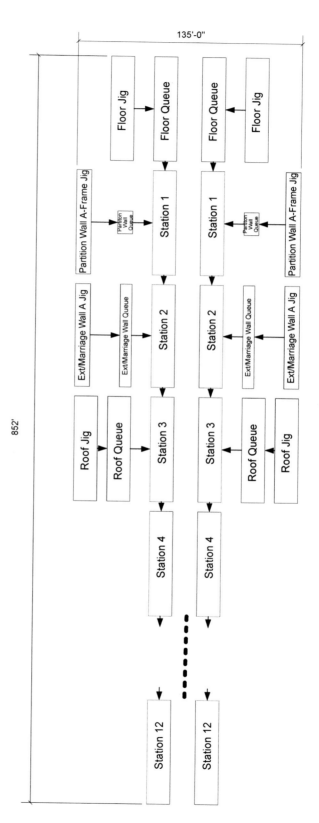

Figure 4.32
Shotgun line layout
with consolidated
queueing for mid-
level production
volume

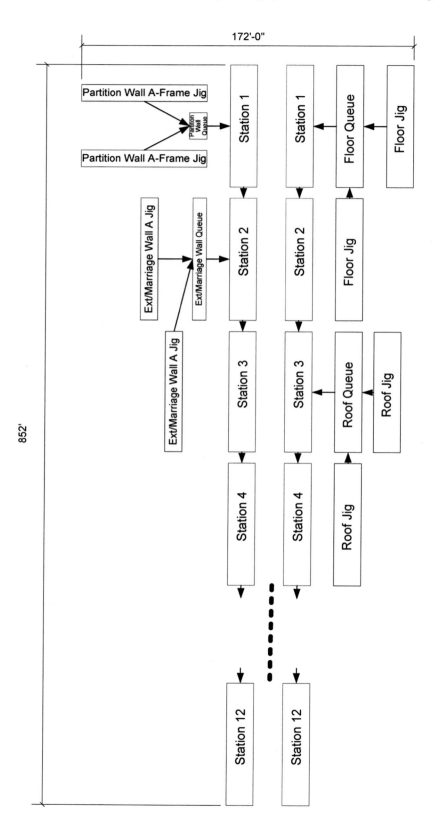

The wide shotgun layout shown in Figures 4.33 and 4.34 improves access to the interior of the module by providing additional floor space between the two lines. This becomes more important as production volume increases. The space can be used to locate workshops (rough plumbing and electric, drywall mud mix, interior door subassembly, cabinet/countertop prep, trim, finish plumbing, finish electric) and to provide staging for materials used on the interior of the module (drywall, interior doors, cabinets, countertops). A mezzanine is incorporated between the two lines to build and queue roofs. The mezzanine can also be used to build and queue subassemblies used on the roof. Partition wall build is located under the mezzanine. The relocation of partition wall and roof build also opens up considerable floor space along the outside of each line. This space can be used for staging materials used on the exterior of the module. The layout has several potential disadvantages. It may require additional floor space. Teams must travel further as they rotate between lines.

As the production volume increases and workstations are added to the lines, the lines can become very long. A T-shaped shotgun layout, such as that shown in Figure 4.35, can be used to reduce the overall length of the factory. Lengthwise orientation should be maintained through the roofing activities that require a mezzanine. A disadvantage of the layout is that teams must travel further as they rotate between lines.

As the production volume increases, TAKT time decreases, and the team size for each activity will increase. Some activities will need to be replicated or decomposed. When an activity located on the line is replicated or decomposed and it lies on a longest sub-path for module-build, a TAKT cycle is added to the duration of the sub-path. An additional workstation is added to each line to accommodate the additional TAKT cycle. Note that each additional workstation can accommodate two activities, since the dwell time at a workstation is twice that of the TAKT cycle. It is important to note that replicating an activity on the line is equivalent to dedicating a team to each line. Therefore, as production volume increases, more activities are replicated, and more teams are dedicated to each line. At some point, it may be advantageous to replicate all activities, dedicate teams to each line and operate the lines independently. Product for each line can then be selected based on specifications such as wet/dry, length, level of finish, or other parameters. TAKT time can be tailored to accommodate line production volume. Activities can be structured to perform the value added work needed. Workstations can be specialized accordingly. Each line becomes a factory within a factory (see Section 4.5.2.10 above). The increased specialization and lower variation on each line should result in higher quality, greater efficiency and less bottlenecking. Other advantages and disadvantages are discussed in Section 4.5.2.10 above.

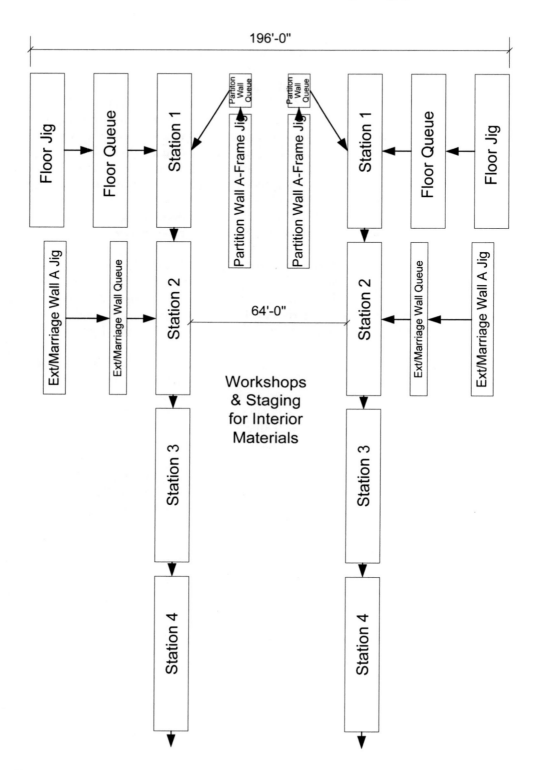

Figure 4.33 Floor level view of subassembly area for wide shotgun line layout for mid-level production volume

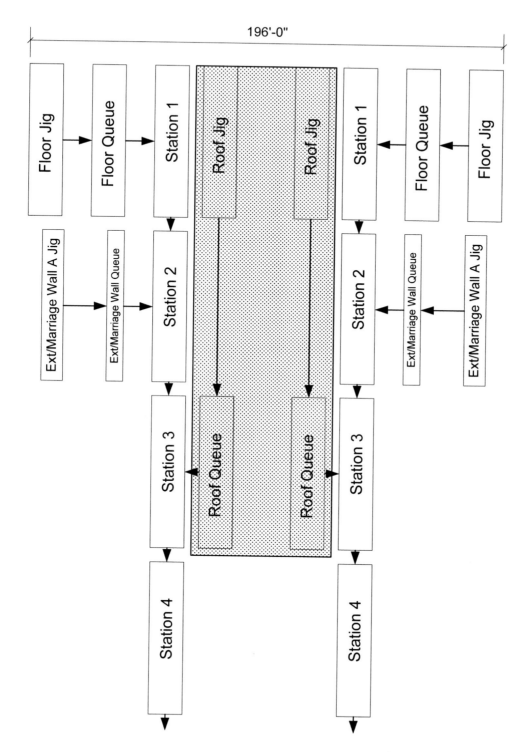

Figure 4.34 Ceiling level view of subassembly area for wide shotgun line layout for mid-level production volume

Figure 4.35
T-shaped shotgun line
layout for mid-level
production volume

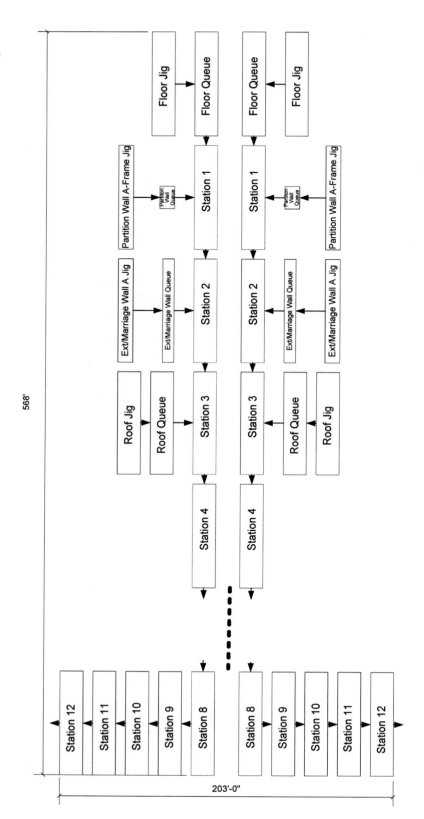

4.6.3 Build-in-place

The build-in-place strategy discussed in Section 4.5.2.15 above can be used to eliminate module movement where it provides little or no value. For example, consider the hybrid sidesaddle/build-in-place layout for high production scenarios shown in Figure 4.36. Modules flow in a sidesaddle orientation through Workstation 17, just beyond the completion of roofing and drywall finishing activities. Each module is then moved to its own build-in-place workstation where it is completed without disruptive module movement. Teams and materials for downstream activities move between the stationary modules.

It is presumed that the progressive assembly line adds value to the activities performed upstream of the build-in-place workstations. These include:

- Build and set major subassemblies – the line allows dedicated cranes to be used to set subassemblies and prevents subassemblies from being moved throughout the factory.
- Complete roofing activities – the line allows fixed mezzanines and catwalks to be used and prevents large, heavy roofing materials from being moved throughout the factory.
- Complete activities on the exterior wall – the line prevents large components such as insulation, sheathing, siding, windows and exterior doors from being moved throughout the factory.
- Complete drywall finishing through sand and paint – the line assures that the modules visited during each production cycle by the unconventional tape and mud team are in close proximity. It also allows the "clean" finish activities to be segregated in a separate area.
- It may be useful to load any remaining large materials (interior doors, cabinets, countertops) into each module before it leaves the line.

Note that modules could be built-in-place much earlier in the process, starting with the setting of major subassemblies or starting after the roof is set.

The build-in-place strategy incorporated in the layout shown in Figure 4.36 has important implications for the facility and the equipment used to move modules. Modules must be moved into and out of each build-in-place workstation. One approach is to extend the line through the build-in-place area to provide access into each workstation. Roll-up doors can be provided at each workstation to allow modules to exit the facility after loading onto a carrier and prepping for shipment.

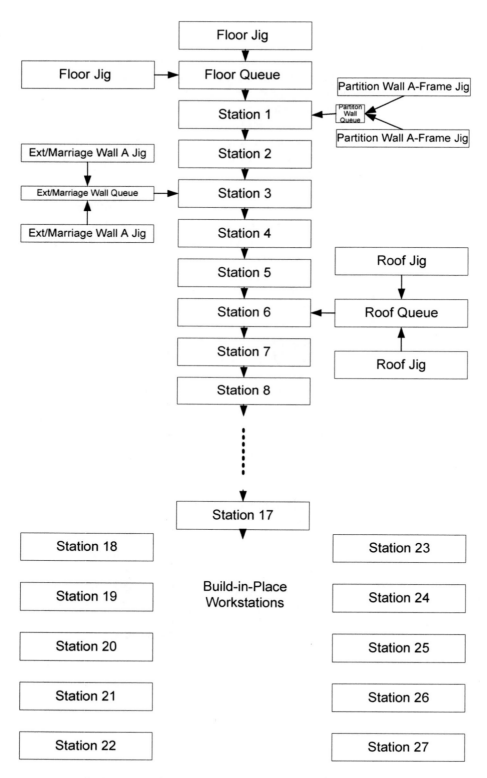

Figure 4.36 Hybrid sidesaddle/build-in-place line layout for high production volume

4.6.4 Roof access

Providing access to the roof is a critical aspect of line design. Roof access is needed to perform the following production activities: setting the roof, installing rough plumbing and electric, insulating, sheathing, shingling, installing fascia and soffit, and prepping/wrapping. The most rudimentary and flexible way to access the roof is by rolling ladder or scaffolding (Figures 4.37–4.38). When activities are performed at specialized workstations on a production line, more substantial and efficient means of access should be considered. Most shotgun production lines use a mezzanine to access the roof. A mezzanine is located on the outside of each line, extending alongside all workstations requiring roof access. The module is oriented with its sidewall facing outward, allowing ready access to the roof from the mezzanine. A mezzanine may also be provided between the two shotgun lines to provide roof access from the marriage wall side of the module. The shotgun lines shown in Figure 4.29 have a mezzanine on the outside of each line and another mezzanine between the lines. The shotgun line shown in Figure 4.39 uses a mezzanine on the sidewall side of the line and a roof supported catwalk on the marriage wall side.

Figure 4.37
Rolling ladders and scaffolding used for upper wall and roof access

Figure 4.38
Rolling scaffold used for
upper wall access

Figure 4.39
Roof supported
catwalk used for
upper wall and
roof access

On a sidesaddle line, roof access is typically provided by narrow catwalks that are located between modules and extend the entire length of the module (Figure 4.20). Access to the catwalks is provided from two mezzanines (one on each side of the line) that extend alongside workstations where access to the roof is required (Figure 4.40). Most catwalks have limited mobility along the line and must be raised to allow module movement on the line. Catwalks may be raised using floor-supported lifts (Figures 4.41–4.42) or using overhead hoists (Figures 4.43). The catwalk shown in Figure 4.43 uses legs for support when it is lowered for operations. Catwalk systems can be designed with three ranges of mobility along the line:

- No mobility – catwalks cannot be moved along the line. Modules must be positioned against a lowered catwalk to provide access. Catwalks are raised using hoists mounted at fixed locations on the factory roof.
- Limited mobility – catwalks have limited mobility along the line, allowing them to be positioned against modules of varying widths. They are raised and positioned using a bridge crane or short monorail cranes.
- Full mobility – a catwalk can be pushed down the line with a module, without being raised. This can greatly reduce line move time. To accomplish this, catwalks roll on rails affixed to the mezzanines on each side of the line (Figure 4.40). When a catwalk reaches the last workstation requiring roof access, it is lifted and transported back upstream using a bridge crane or long monorail cranes.

Figure 4.40
Mezzanine alongside sidesaddle line (left) providing access to catwalks spanning line

Figure 4.41
Catwalk on floor-supported lifts spanning sidesaddle line

Figure 4.42
Catwalk on floor supported lifts

Figure 4.43
Catwalk spanning
sidesaddle line (note that
it is raised using overhead
hoists and that legs are
used for support)

The mezzanine must be sufficiently wide to allow workers carrying tools and materials to pass. If materials, tools and equipment are to be staged on the mezzanine, sufficient space must also be provided on the outside of the mezzanine. A mezzanine used for roof access can be integrated with a larger mezzanine used for material storage or production activities (such as roof build shown in Figure 4.24). Although the floor space under a mezzanine used for roof access is limited, it is extremely valuable, since it fronts workstations on the production line. Material staging, tools and workshops are often located here (Figures 4.44, 4.45).

Figure 4.44
Mezzanine alongside shotgun line houses long framing table (background) and queue for side and marriage walls (note workshop and materials located underneath)

Figure 4.45
Mezzanine alongside sidesaddle line with workshop and materials located underneath

The location of mezzanines and catwalks can impair production performance. For example, the "Prep/Drop Roof & Wrap for Shipment" activity might ideally be performed near the end of the line. This minimizes the impact of wrapping materials on exterior wall activities and on module entry from the marriage wall side. However, activities that require roof access are often located at adjacent workstations (earlier in the process) to simplify mezzanine configuration. If a workstation is equipped to perform multiple activities, a mezzanine required by one activity can impair material flow for another activity. For example, consider the line layout shown in Figure 4.22. To improve worker mobility, assume that Workstation 3 must accommodate both exterior/marriage wall and roof set. If the roof access mezzanine extends to Workstation 3 (to support roof set), then exterior/marriage walls must be lifted, either over the mezzanine or over the modules in workstations two and three.

Worker safety is a critical issue for activities that are performed on the roof. Ideally, most of these activities would be performed before the roof is set on the module. However, the need to integrate services in the wall and roof, before insulating and roofing, has prevented producers from achieving this milestone. Several measures can reduce the risk of worker falls. Mezzanine height should provide safe and easy access to the roof, regardless of wall height. Safety rails should be provided along the perimeter of the mezzanine wherever frequent access is not required. Gates can be provided in the safety rails where infrequent access is needed (for example, at material staging locations along the back side of the mezzanine). Stairs equipped with hand rails should be used to access each mezzanine. Shotgun lines must accommodate modules of varying widths and overhang configurations. Therefore, it may not be possible to position every module on the line so that it is easily accessible from the mezzanine. A slide-out catwalk can be integrated on the inside of the mezzanine to improve access to the roof (Figure 4.46). Tethers and nets can reduce the likelihood of injury should a worker fall from the roof. The factory shown in Figure 4.44 illustrates many of these safety features: a safety net is stretched between modules; safety rails are provided along the perimeter of the mezzanines; and slide-out catwalks extend from underneath the mezzanine.

Figure 4.46
Mezzanine alongside shotgun line (left) with slide-out catwalk (overhead)

4.6.5 Material handling for modules

The module is the largest, heaviest and most difficult component that is handled in the modular factory. The most rudimentary and flexible way to move a module between workstations on the line is on castors. Casters are attached to the underside of the module floor prior to placement on the line (Figure 4.47). A module on casters can be moved in any direction, even 90 degree turns. No tracks are required on the factory floor, which facilitates the flow of traffic across the line. There are disadvantages when using casters. Heavy modules can quickly wear groves in a concrete factory floor. Steel plates are sometimes placed on the factory floor along the flow path to prevent wear. It is difficult to move a module on casters in a straight line or to make controlled turns. This makes movement and positioning along the line more difficult. Figure 4.48 illustrates the use of casters rolling on a steel plate. Note that the plate is formed with an inverted V-shaped guide that forces the module to roll on a straight line.

Figure 4.47
Floor with casters attached and supported by stanchions to allow access from below

Figure 4.48
Module with casters rolling on steel floor plate with inverted V-shaped guide

A more common approach is to move modules along the line on parallel rails mounted on the factory floor (Figure 4.49). Multi-wheeled bogies mounted to the underside of the module floor ride on the rails. Ideally, the rail profile is low enough to allow traffic to move easily across the line. A similar approach is to permanently mount rollers on the rail, creating long narrow roller conveyors. Steel skids mounted to the underside of the module floor ride on the conveyors (Figure 4.50). Rails are sometimes located in a groove formed in the factory floor to reduce disruption of cross traffic. However, the groove can accumulate construction debris, which can increase housekeeping and hinder line movement. Reorienting a module on the line or moving a module through a turn when using rails requires additional equipment and effort. Reorienting a module on a straight line involves lifting the module off the rails, rotating it 90° on its axis, reorienting the wheels, repositioning the module and lowering it onto the new rails. Moving a module to a perpendicular line while maintaining the same orientation involves lifting the module off the rails, reorienting the wheels, repositioning the module and lowering it onto the new rails. Moving a module to an adjacent parallel line while maintaining the same orientation involves lifting the module off the rails, repositioning and lowering it onto the new rails. These moves can be accomplished using a high-capacity bridge crane, floor jacks/casters or a turntable/transfer car embedded in the floor.

Figure 4.49
Module
movement on
rails

Figure 4.50
Module
movement on
narrow roller
conveyor

Module propulsion along the line can take many forms. Workers commonly push modules along the line (Figure 4.51). However, pushing a module becomes more difficult as weight is added, requiring more workers. Riding lawn tractors or specialty pushers (Figure 4.52) are also used. Some producers use powered chains to pull modules along the line.

Figure 4.51
Workers pushing
module along line

Figure 4.52
Specialty equipment for
pushing module along
line

More sophisticated, flexible and expensive methods can also be used to move modules along the line. Modules can be moved on air casters that float the module on a thin cushion of air (Figure 4.53). Modules on air casters can easily be moved and positioned by workers. Reorientation and turns are easily navigable. Although no rails are needed, air casters require hoses for compressed air. Automated transport alternatives such as powered conveyors, automated guided vehicles and automated electrified monorails may also be used to move modules. However, they have not been practical due to module size and weight, the need for easy module access during the build process and system cost.

Figure 4.53
Air casters used for module
movement along line

A heavy duty overhead bridge crane or a mobile gantry crane can be used to move modules without rails. The gantry crane provides the additional flexibility to move a module between crane bays and even outside the factory. Therefore, it might be used to service build-in-place workstations or to stage a finished module that is waiting for shipment without dedicating a carrier.

Queued modules typically remain on the line, where they can be easily accessed by their next activity. Finished modules awaiting shipment are usually staged in the yard on their carriers. To conserve carriers, some producers stage finished modules on blocks and delay loading the module onto the carrier until shipping is imminent. Special purpose off-line storage equipment is not used due to module size and weight, the need for ready access and system cost.

4.6.6 Material handling for major subassemblies

Major subassemblies (floor, walls, roof, ceiling) are also large, heavy and difficult components to handle. To minimize handling difficulty, subassemblies should maintain the same orientation throughout the process, including building, moving, storage and set. Fortunately, with the exception of the roof, the subassemblies are two-dimensional. This allows some flexibility in orientation during building, movement and storage, including the ability to stack.

In the modular factory, subassemblies are usually built-in-place, flat on a framing table. Walls can also be built vertically on an A-frame jig. With the possible exception of the roof, subassemblies can also be built progressively, queued and moved flat on a roller conveyor.

After being built, a wall is usually stored vertically, standing on its bottom plate. This maintains orientation, conserves floor space and prevents stacking, which can lead to drywall damage and restacking (to access the next wall during wall set). Posts or rails are used to support walls when they are stored vertically. Partition walls are the smallest and lightest of the major subassemblies, due to their shorter length and smaller (2" x 4") framing lumber. They are often stored and moved to the line on carts (Figure 4.54). They can also be stored on the factory floor (Figure 4.55). Handling to and from storage is usually by manually lifting/dragging the wall on its bottom plate. If a crane is used, partition walls are lifted by the top plate, although they are sufficiently rigid to be lifted flat.

Figure 4.54
Cart used to store
partition and end walls

Figure 4.55
Stand used to support partition walls stored on
factory floor

Larger and heavier walls are usually queued on the factory floor (Figure 4.44, 4.56). They may also be stored in special storage equipment designed to accomplish other objectives. For example, the elevated storage rack shown in Figure 4.57 allows walls to be stored with wall-building materials staged underneath. The mobile storage rack shown in Figure 4.58 is a large, unpowered transfer car that travels on rails mounted on the factory floor. The rack's mobility allows walls to be placed in the rack using one overhead crane system, remain in storage until needed, moved a short distance and then removed by a second, perpendicular crane system for transport to the line. Large and heavy walls are usually moved by overhead crane, lifting from the top plate. Long walls are not sufficiently rigid to be lifted flat.

Figure 4.56
A-frame stand (left) used to support side and marriage walls stored on factory floor

Figure 4.57
Elevated storage rack (top) used to store side and marriage walls (note building materials staged underneath)

Figure 4.58
Mobile storage rack for
side and marriage walls
allows access by two
perpendicular crane
systems

The floor is larger and heavier than walls. It is usually stored flat on the factory floor because of its size and weight and to maintain orientation. Although floors can be stacked to provide additional storage capacity, this can lead to restacking and damage to vinyl flooring. Movement from the floor jig to storage is usually by overhead crane. The floor is usually lifted flat to maintain orientation, although it can be lifted by one edge of the rim joists and stored vertically on the opposite edge. If casters are used for module handling and the storage queue is located adjacent to the first workstation on the line, then the casters may be installed on the floor while it is in the queue. This allows the floor to be pushed from the queue to the first workstation. The same is true when modules are handled on rails, except that the queue must be located in-line with the first workstation. Figure 4.59 shows a floor queue located at the beginning of a rail line. Note that the bottom floor already has the wheels attached and can be rolled to the first workstation. Note also that two additional floors have been stacked on the first, requiring that they be unstacked before the bottom floor is accessed.

The roof is the largest, heaviest and most difficult major subassemblies to handle. The roof is three-dimensional. Therefore, it must be stored flat and cannot be stacked. Drywall is installed on the bottom surface of the roof and ceiling, requiring special storage equipment to prevent damage. For example, the queued roof shown in the background of Figure 4.60 is supported on stanchions. A queued roof may also be hung on chains fixed to the factory ceiling. Elevating the queued roof above the factory floor provides valuable space underneath for material staging, saws and other tools and equipment. It also allows rough electric installation to begin in the roof, beginning with the installation of electrical boxes. The roof is moved by overhead crane in the flat orientation.

Figure 4.59
Stacked queue of floors
located at beginning of
rail line

Figure 4.60
Roof queued on
stanchions
behind framing
table (note
materials staged
underneath
queue)

The ceiling is two-dimensional and has more handling flexibility than the roof. However, it is built on the same jig and set on the module at the same workstation on the line as the roof. Therefore, it is practical for the roof and ceiling to use the same handling equipment and procedures.

4.6.6.1 Overhead cranes

Overhead cranes play the primary role in handling most major subassemblies. They can also be used to move unit loads of building materials to hard-to-access staging locations in the workplace and even to distribute large, heavy materials directly to their point-of-use in difficult-to-reach locations (for example, bringing decking and shingles directly to workers on the roof). Cranes can also be used to support production activities such as lifting folding roof elements or loading a module onto a carrier. Providing adequate crane capacity is crucial to factory performance. Inadequate crane capacity delays movement of subassemblies and other materials to the line. This can delay the completion of set activities, which disrupts flow on the line, creates idle time and reduces production capacity.

Two types of overhead cranes are commonly used in the modular factory: monorail cranes and bridge cranes. The simplest monorail crane consists of a trolley (mechanical or powered) that traverses a single overhead rail mounted on a structural beam. A hoist (mechanical or powered) is attached to the trolley to lift/lower a load. Major subassemblies can be moved lengthwise along the rail, typically carried by several trolleys. Longer/wider subassemblies such as floors, ceilings and roofs often are picked up at multiple points along their length and width in order to distribute the load more evenly. This is accomplished using a long, rigid spreader bar that is hung from the hoists and attached to the load at multiple points. The single monorail crane system is ideal for moving subassemblies lengthwise between any two points along a straight line (under the rail). Several parallel monorail cranes can be used jointly to move a subassembly widthwise, with the same limitation to linear movement.

The sidesaddle line shown in Figure 4.61 uses monorail cranes to move the floor, exterior/marriage walls and the roof. Multiple monorail cranes are used to move the floor widthwise from the jig to the queue. An independent monorail crane running perpendicular to these cranes is used to move the floor lengthwise from the queue to the first workstation on the line where partition walls are set. The cranes are located as close as possible to each other over the queue so that the queue can be accessed by all cranes, safely and without damage. Similar monorail cranes are used to handle exterior/marriage walls and the roof. Note that the alignment of subassembly workstations allows the monorail cranes that service the floor and roof jigs to be consolidated, simplifying the system and adding flexibility. The cranes serving both sides of the A-frame exterior/marriage wall jig are also consolidated. A similar crane system is shown in Figure 4.62. Two monorail cranes (background) are being used to move a floor from the jig (left) to a queue/workstation (right), where vinyl flooring is installed, if needed. A spreader bar is being used to distribute the load. A single monorail crane (right side) is used to move a floor from the queue (background) to the first workstation on the line (foreground). A spreader bar is hanging from the crane. The crane rail in the center is used to move exterior/marriage walls from a queue (in foreground – not shown) to the second workstation on the main line, where they are set on the module.

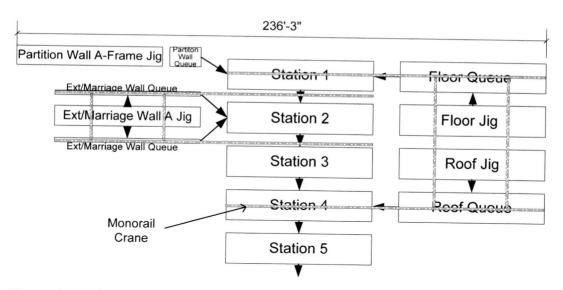

Figure 4.61 Subassembly area of sidesaddle line layout served by monorail crane system

Figure 4.62
Monorail cranes
used to move
floor and walls

A bridge crane consists of two primary elements: a bridge and a runway. The bridge is analogous to a single monorail crane. It consists of long single or double girders (beams), a rail supported by the girders and a trolley that travels along the rail. Unlike a monorail crane, the bridge is mobile. An end truck is attached to each end of the bridge, allowing it to ride on rails that run perpendicular to the bridge. The runway consists of these rails and their supporting beams, brackets and framework. A bridge crane can move a load between any two points in the crane aisle, defined by the span of the bridge and the length of the runway. Multiple bridges can use the same runway, allowing the bridge crane system to serve multiple functions at the same time along the length of the crane aisle.

A monorail crane offers several advantages over a bridge crane. There are fewer components, and, therefore, it is less complex and less expensive to purchase and maintain. The monorail crane has less impact on the facility. Because it does not require a bridge, very little overhead clearance is needed. A monorail crane system can even be used under a mezzanine, as demonstrated by the three monorail cranes used to move floors in Figure 4.63. No column can lie in the travel path of a load carried by a crane. This can be particularly restrictive for a bridge crane, with its wider crane aisle. Therefore, use of a bridge crane can constrain factory structure and increase facility costs.

Figure 4.63
Floor build activity supported by three monorail cranes located under a mezzanine

The greatest disadvantage of a monorail crane is that it can only move a load in a straight line. This limits its ability to precisely position a load, interface with other crane systems and serve as part of a broader material handling system. The crane configuration shown in Figure 4.61 performs well when moving the floor, since the origin and destination lie under the rails for both moves. However, the configuration does restrict flexibility, limiting the queue to a single floor (without extraordinary efforts) and requiring that the floor be dropped in the queue and reloaded onto the set crane, even when it is needed immediately on the line.

The monorail crane configuration does not perform as well when moving exterior/marriage walls. As the wall queue grows wider, it is increasingly difficult for the cranes to access all storage positions in the queue. This can occur when queue size is increased or two physical queues are consolidated. When the cranes cannot access all storage positions in the queue, the queue itself might be moved for easier access by both cranes. The mobile storage rack shown in Figure 4.58 rolls on floor-mounted rails, allowing queued exterior/marriage walls to be accessed by three cranes: a bridge crane that moves walls from the jig to the queue and two monorail cranes that move walls from the queue to two locations on the line. Another approach is to use a stationary wall queue with a dedicated pickup position. The next wall to be set is moved from its original storage position in the queue to the dedicated pickup position that is accessible by all cranes. This requires a second handling of each wall by the cranes that moved it to the queue.

Other issues affect the performance of this monorail crane configuration when moving exterior/marriage walls. Setting the wall requires precise positioning of the wall on the module. Since a wall on a monorail crane cannot be moved perpendicular to the crane rail, the module must come to the wall. This requires two precision moves for the module, one to set the sidewall and another to set the marriage wall. The monorail crane configuration also limits the flexibility of the production system to accommodate varying cycle times. A wall must be dropped in the queue and reloaded onto the set crane, even when it is needed immediately on the line. More importantly, wall set is constrained to workstations accessible by the two cranes used to set the walls. This includes:

- Workstation 1 – only the leading wall can be set.
- Workstation 2 – the preferred workstation for setting exterior and marriage walls. Both leading and trailing walls can be set (with some repositioning).
- Workstation 3 – only the trailing wall can be set.

This restricts the mobility of the wall set team, limiting their movement to neighboring workstations when cycle times vary.

The monorail crane configuration has similar limitations on roof movement. Roof set requires precise positioning of the roof on the module, necessitating a precision move of the module. Queue size is limited to a single roof, and the roof must be dropped in the queue and reloaded onto the set crane, even if it is needed immediately on the line. Roof set is constrained to a single workstation (Workstation 4). Completing roof set during its scheduled TAKT cycle is a continuing problem for many modular producers. Delays in building and setting the walls can delay the start of roof set. Roof build can also delay the start of roof set. Delays in roof build are often caused by roof complexity, which contributes to extended cycle times, rework and design/build questions that must be resolved. If a monorail crane is used, roof set is constrained to a single workstation on the line. This allows a delay at roof set to bottleneck upstream workstations and create a vacancy downstream on the line. This results in idle workers, diminished capacity and greater difficulty in recovering.

A more flexible crane configuration can be created using both monorail and bridge cranes. For example, the crane configuration shown in Figure 4.64 reflects the following changes:

- The monorail cranes used to move subassemblies from the queues to the set workstations on the line are shortened, terminating before they reach the line.
- A new bridge crane is added to serve the line.
- Computer controlled transfer points are added at the end of each shortened monorail crane, allowing trolleys to transfer between the monorail and a bridge, which can be positioned and locked at the transfer point on the crane runway during transfer.

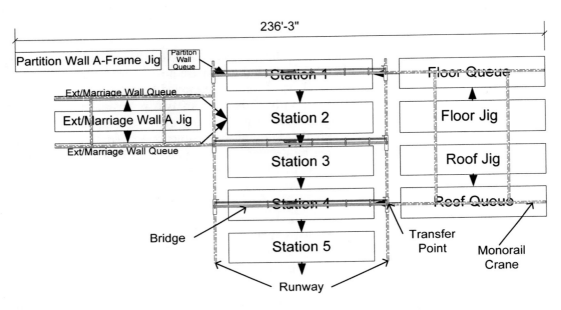

Figure 4.64 Subassembly area of sidesaddle line layout served by monorail/bridge crane system

The bridge crane spans the width of the line, and its runway runs the length of the line. Three bridges are used, one each to set the floor, walls and roof. Each bridge is capable of precisely positioning a subassembly on a module located at any workstation on the line. This allows the walls and roof to be set without precisely locating the module. It also gives set teams the mobility to perform their work anywhere on the line. It should be noted that the monorail crane configuration over the jigs and queues has not changed. Therefore, queue size is still limited and a subassembly must still be dropped into a queue, even if it is needed immediately on the line. Several minor changes to the configuration might also be considered:

- Set teams may be able to share a bridge, reducing the number of bridges required on the runway. For example, the same bridge might be used to move the floor and set the walls. This might be facilitated by the use of lean setup time reduction techniques such as installing wheels on the floor while it is in the queue, rather than while supported by the crane.

- The floor might be moved from the queue to the first workstation using a monorail crane such as that shown in Figure 4.61. The bridge crane would be shortened to prevent interference with the longer monorail crane used to move the floor. This would eliminate the need for a bridge dedicated to the floor.

Additional flexibility can be gained by replacing the monorail cranes in each subassembly bay with a bridge crane. For example, the crane configuration shown in Figure 4.65 reflects this approach. Note that the subassembly area is reconfigured so that subassembly operations requiring a crane are located on one side of the line. This allows one bridge crane to handle all movement of exterior/marriage walls and roofs from their jigs to the bridge crane serving the line. Also note that a single monorail crane is used to move the floor from the jig to the queue, eliminating the potential need to dedicate two bridges to the floor. Wheels are installed on the floor as it is lowered into the queue, allowing the floor to be rolled from the queue to the first workstation. This crane configuration provides the added flexibility to move freely over the wall queue, eliminating wall queue size as an issue. As a result, the two exterior/marriage wall queues are consolidated. Exterior/marriage walls and the roof can also be moved directly from the jig to the line, without being dropped into a queue, when they are needed immediately.

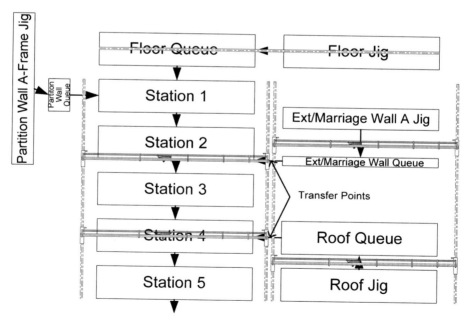

Figure 4.65 Subassembly area of sidesaddle line layout served by bridge crane system

Bridge cranes can also be oriented with the bridges parallel to the line, so that a single crane can serve all activities involving a subassembly, from jig to set. This crane configuration can be effective for high production scenarios where subassembly jigs are replicated. The crane configuration shown in Figure 4.66 is an example of this approach. One bridge crane serves the exterior/marriage wall jigs, the queue, the wall set workstation on the line (Workstation 3) and the two neighboring workstations. The second bridge crane serves the roof jigs, the queue, the roof set workstation (Workstation 6) and the two neighboring workstations. Floors are moved from the jigs to the queue using a monorail crane for lengthwise moves and two monorail cranes for widthwise moves.

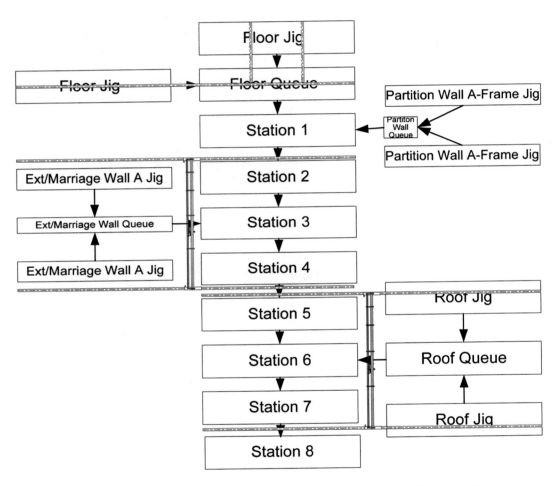

Figure 4.66 Subassembly area of sidesaddle line layout for high production volume served by bridge crane system

The crane configuration shown in Figure 4.66 provides similar capabilities to those of the configuration shown in Figure 4.65. The primary operational advantage is that a subassembly is not transferred between bridge cranes when moving from the queue to the line. There are several operational disadvantages. Exterior/marriage walls and the roof can only be set in a limited number of workstations. Although it is possible to overlap exterior/marriage wall set and roof set workstations by linking these systems using transfer points, it is likely to cause conflicts between these activities and is not recommended. Because the subassembly is oriented perpendicular to the bridge, only one trolley/hoist per bridge carries the subassembly. Moving a large subassembly using only one hoist requires precise positioning for balance. Two interlocked bridges with interlocked trolleys/hoists can be used to address the balance issue; however, this doubles the number of bridges required. Another approach to the balance issue is to reorient the runway and bridge 90 degrees over the same crane aisle. This results in a short runway that runs parallel to the line over three workstations and a long bridge that spans the width of both the line and the subassembly area. The configuration allows a subassembly to be supported by multiple trolleys/hoists on the long bridge.

The crane configuration shown in Figure 4.66 also has design and capital cost advantages. A shorter bridge span is required. Crane components can be sized for the specific load to be handled, rather than the heaviest load (typically the roof). It should be noted that this configuration can be effective for high production scenarios where subassembly jigs are replicated. There are several reasons:

- In high production scenarios, the length of the subassembly area that needs to be served by a crane is often equal to the length of the line that needs to be served. For example, in Figure 4.66 each crane serves three positions in the subassembly area: two subassembly jigs and one queue. The crane also serves three workstations on the line: the preferred set workstation and its two neighboring workstations (allowing some mobility).
- In high production scenarios, the workstations where major subassemblies are set need not overlap, because of queues placed in the line. For example, in Figure 4.66 workstations four and five are used as queues between exterior/marriage wall set and roof set. This allows workstations two through four to be dedicated to exterior/marriage wall set and workstations five through seven to be dedicated to roof set.

Similar crane configurations are used for shotgun line layouts. Parallel monorail cranes can be used to move a subassembly widthwise from the jig to the queue and from the queue to the line where it is set. This crane configuration requires precise positioning of the module for set. However, it does provide some flexibility. It can access both of the side-by-side workstations where the subassembly is set. The cranes can also move the subassembly directly from the jig to the set workstation, if it is needed immediately. A hybrid crane configuration using monorail and bridge cranes can add flexibility. Monorail cranes can be used to move a subassembly from the jig to the queue and from the queue to a transfer point, where the load is transferred to a bridge crane serving the lines. This hybrid crane configuration allows a subassembly to be precisely positioned on a module located at any workstation on the lines.

If a subassembly queue is oriented lengthwise alongside a shotgun line, it is possible to overlap monorail cranes with a bridge crane, allowing both crane systems to access the queue. For example, the wall queue shown in Figure 4.67 is located on a mezzanine alongside a shotgun line. Multiple monorail cranes are used to move a wall widthwise from a jig (located behind the queue on the mezzanine) to the queue. The monorail crane rails extend over the full width of the queue, allowing the cranes to access all storage locations in the queue. A large bridge crane is used to move the wall from the queue to the line where it is set. To allow the bridge crane to access the queue, a runway rail is located over and perpendicular to the monorail rails. The bridge is able to move on the runway rail above the monorail rails because the supports for the monorail rails, which are hung from the ceiling of the facility, are positioned inboard of the runway rail. Note that since the crane aisles overlap, conflicts between the crane systems are still possible. To avoid potential conflicts, the two crane systems should not access the queue at the same time. Also, the bridge cannot move over the monorail rails until its load clears the rails.

Figure 4.67
Overlapping
monorail and
bridge cranes
allow both to
access queue
for side and
marriage walls

Although a crane configuration utilizing only bridge cranes could be used for a shotgun line layout, it provides few additional advantages.

Bridge cranes are typically oriented so that the bridge spans the shorter dimension of the crane aisle. This minimizes design load factors and often provides greater operating flexibility if multiple bridges are needed. However, it is possible for the bridge to span the longer dimension. This can allow a crane to better meet special handling needs, such as the case discussed previously in this section (see Figure 4.66). In this scenario a subassembly is handled in an orientation perpendicular to the bridge, creating a balance problem. Reorienting the bridge to span the long dimension of the crane aisle would allow the subassembly to be oriented parallel to the bridge. This solves the balance problem by providing support from multiple trolleys/hoists on the bridge. Intermediate runway rails and trucks will likely be required to help carry the longer bridge. The bridge crane shown in Figure 4.68 provides an example of this approach for mid-level capacity shotgun lines. The crane serves two end-to-end roof jigs (foreground) and two adjacent roof set workstations on each of the two shotgun lines (background).

Figure 4.68
Bridge crane
with short
runway and
wide span
used to move
the roof

It can be difficult to assess the capacity of a complex crane system in a modular factory. Scenarios that should be evaluated carefully include:

- Crane systems where the same hardware handles multiple functions – for example, a bridge crane where the same bridge handles movement of walls and the roof.
- Bridge cranes where the routine travel paths of multiple bridges overlap.
- Integrated crane systems that use a transfer point to transfer loads between cranes.
- Any crane subject to high cycle time variation – for example, using a crane to move partition walls.

Inadequate crane capacity delays movement of subassemblies and other materials to the line. This can delay the completion of set activities, which disrupts flow on the line, creates idle time and reduces production capacity.

The design of the crane system affects the design of the factory structure, impacting building columns and the structural design and clear height of the factory roof. Cranes require substantial structural support. Providing this support is complicated since no column supporting the factory roof or crane can be located within the operating area of the crane. The operating area can be defined as the travel footprint of any load carried by the crane. A load can extend well beyond the crane itself. To accommodate an overhanging load, a monorail crane is typically supported from above. A bridge crane runway can also be supported from above or from columns. If columns are used to support a bridge crane runway, the operating area must lie inside the perimeter of the crane aisle defined by the crane rails. If the bridge crane runway is supported from above, the bridge can overhang the runway to enlarge the crane aisle and operating area (see Figure 4.69). Overhead support for a crane is usually provided from the factory roof. However, it may also be provided by an independent structure whose support columns lay outside the operating area. It is often possible to reduce total facility costs by integrating the crane support structure with the facility.

Figure 4.69
Bridge crane
serving roof jig
(left) and roof
queue (right)
showing bridge
overhanging
runway

The clear height of the factory roof (from the factory floor to the lowest overhead obstruction) is often driven by the design of the crane system. Crane design factors include the minimum height required for the runway, trucks, bridge, trolley, hoist and spreader bar. Equipment and activities that are located under the crane system must also be considered. If adequate clearance is not provided, operations under the crane can be awkward, suffering from inefficiencies and even interruptions when the crane is being used. For example, consider the overlapping monorail/bridge crane configuration shown in Figure 4.67 that was discussed above. The configuration allows both cranes to access a queue of walls that is located on a mezzanine. Note, however, that overhead clearance is insufficient to lift a wall over another wall in the queue. Therefore, a wall must enter the queue from one side (placed in the last accessible empty storage location) and exit the queue from the other side (retrieved from the first, and only, accessible full location). Over time, the occupied storage locations shift toward the entry side of the queue. When an empty storage location is no longer accessible for an entering wall, all walls in the queue must be relocated to the other side of the queue (nearest the line), freeing storage locations for entering walls.

Roof-related activities (build, queue, set, install rough plumbing/electric, insulate, sheath, shingle, install fascia/soffit, wrap) that are located under a crane must be carefully considered when determining clear height. Some activities may be located on a mezzanine (roof build, roof queue). Some activities can only be performed (or are more efficient) when the roof is extended. Some activities may require another roof to be moved overhead. For example, the monorail crane configuration shown in Figure 4.61 requires the roof to be set in Workstation 4. The roof does not pass over another roof. The crane configurations shown in Figures 4.64 and 4.65 allow the roof to be set anywhere on the line. Workstation 4 is designated as the preferred set workstation and is used most often. When a roof is set in Workstation 4 or further downstream, it does not pass over another roof. However, if a roof is set in Workstation 3 or further upstream, then it must be lifted over at least one roof that has already been set. If this previously set roof is extended, then the roof being set must be lifted even higher. If adequate clearance is not provided, roof set may be delayed. Note that this constraint can be relieved somewhat by reversing the location of the roof jig and the queue/transfer point on the layout. This allows a roof to be set in Workstation 3 without passing over another roof that has already been set. A disadvantage of relocating the transfer point is that any roof that is set downstream of Workstation 3 must be moved further after being transferred to the bridge crane serving the line.

Shotgun line layouts face similar overhead clearance issues. However, the parallel lines create another obstruction, passing over one line to access the other. Producing the roof between or on both sides of the shotgun lines can help resolve this issue. Figure 4.68 demonstrates many of the elements that must be considered when determining clear height: a roof jig located on a mezzanine, a bridge crane that transports completed roofs to roof set and roofs on the production line at various stages of unfolding.

4.7 DESIGN THE FACTORY FACILITY

In this phase the factory facility is configured around the production line. Primary features of the facility are planned. Remaining production support areas are located. Items to be considered include:

- Workstations for minor subassemblies – mill, window/door openings, subassemblies for roof, rough plumbing, finish plumbing, interior doors, shiploose subassemblies
- Workshops for line activities – electrical, plumbing, drywall mud mix, molding, shiploose
- Material storage areas – stockrooms, warehouses
- Personnel support areas – production offices, QC offices, first aid, break room, bathrooms
- Electrical/mechanical support areas – electrical panel, air compressors, backup generator, drywall mud separator, corrugated compactor
- Primary facility features – walls, columns, mezzanines, doors, docks, clear height
- Surrounding yard storage and flows

Locating workshops and material storage areas that directly support key activities on the line is of special importance during this phase. For example, it is not unusual to find the drywall mud mix room, the electrical stockroom and the insulation storage area far from their associated activities on the line. This results in lengthy trips to resupply materials. When replenishment is not well planned, frequent unexpected trips to retrieve additional materials or special tools can greatly reduce production efficiency, add to cycle time variation and contribute to flow disruptions throughout the factory.

Other corporate activities are often located on the same site as the factory. These include engineering and sales offices, model homes and a carrier refurbishment facility. Although these activities could be located within the factory facility, they are most often located in separate structures and do not affect factory operation. Therefore, they are not considered in this discussion.

The graph shown in Figure 4.70 shows the relationship between overall plant size and production capacity for modular producers. It shows the expected increase in plant size associated with an increase in production capacity. However, it also shows considerable variation. Likely causes of the variation include line length and layout (see Section 4.6 above), available facility size and capital availability.

Figure 4.70
Relationship between
plant size and
production capacity
for modular producers
[9]

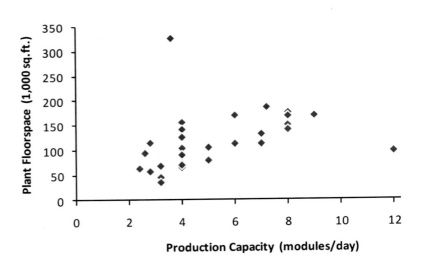

An overall factory design around a sidesaddle line is shown in Figure 4.71. The area of the primary facility is roughly 100,000 square feet, excluding adjacent personnel and electrical/mechanical equipment support areas. Workstations for minor subassemblies, workshops for line activities and material storage are located inside the facility in designated areas close to their point-of-use. The column layout features a central 75' wide north-south bay to accommodate sidesaddle module flow on the line. A bridge crane with two bridges serves all workstations in this bay. Access to the module roof is provided by eight catwalks that span the bay. The catwalks serve the line workstations that are preferred for roof-related activities. Access to the catwalks is provided by a mezzanine on each side of the line. Access to the roof in other stations is provided by rolling ladders or scaffolding. A second 75' bay is located on the east side of line to accommodate major subassembly flow from jig to queue. A shorter bridge crane with two bridges serves subassembly activities in this bay. A monorail crane is used to move the floor from the jig to the queue. A shorter column spacing may be used elsewhere in this bay where the wide bay is not needed. A 50' bay is located to the west of the line to accommodate various subassembly workstations, support workshops and material storage. A 20' bay is provided on the east and west perimeter of the building for a perimeter aisle and material storage against the factory walls. North-south column spacing is 25', except for the north and south perimeters of the building where it is extended to accommodate the perimeter aisle and storage against the walls. The column layout may be changed by the construction contractor during final facility design after considering structural engineering and financial factors.

No large scale mezzanines are provided for subassembly production or storage. Roll-up doors are located on the east and west walls for material delivery and module movement into and out of the building. The floor of the factory is at-grade, allowing all vehicles to enter and leave the building. Most building materials are delivered on flat bed trailers. These materials are unloaded outside the building using a forklift truck operating from the ground. The forklift then transports the materials into the factory through a nearby door and stores them in a designated area. Smaller materials and those susceptible to weather damage are delivered in a conventional enclosed trailer. These trailers are accommodated at two conventional loading docks equipped with dock levelers on the west side of the building. Workers unload the trailer using a pushcart or a low-profile forklift truck. Although not explicitly shown on the layout, it is assumed that some modules will be staged temporarily around the building while awaiting completion and/or shipping. The site must accommodate the movement and staging of these modules as well as delivery trucks, employee parking, and other structures required by the company.

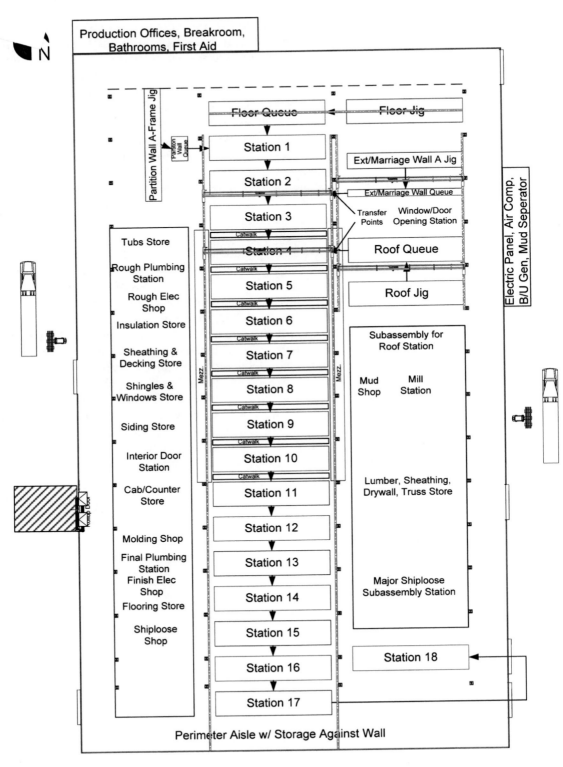

Figure 4.71 Overall factory design with sidesaddle line layout

An overall factory design around a J-shaped sidesaddle line for high production scenarios is shown in Figure 4.72. The area of the primary facility is roughly 143,000 square feet, excluding adjacent personnel and electrical/mechanical equipment support areas. Workstations for minor subassemblies, workshops for line activities, and material storage are located inside the facility in designated areas close to their point-of-use. These areas are increased by roughly 50% to accommodate the high production volume. The column layout features a central 75' wide north-south bay to accommodate sidesaddle module flow on the first half of the J-shaped line. A bridge crane with two bridges serves all workstations in this bay. Access to the module roof is provided by 13 catwalks that span the bay. The catwalks serve the line workstations that are preferred for roof-related activities, plus one workstation on each end to allow worker mobility. Access to the catwalks is provided by a mezzanine on each side of the line. Access to the roof in other stations is provided by rolling ladders or scaffolding. An adjacent 75' bay to the west accommodates module flow on the second half of the line. It also houses exterior/marriage wall subassembly, a floor jig and three workstations used to customize select modules. A bridge crane serves wall subassembly operations in this bay. A monorail crane is used to move the floor from the jig in this bay to the queue. A third 75' bay to the east of the line accommodates partition wall and roof subassembly. A bridge crane serves roof subassembly operations in this bay. A shorter column spacing may be used elsewhere in this bay where it is not needed. A fourth 75' bay is located to the west of the line to accommodate various subassembly workstations, support workshops and material storage. Other features of the design are similar to those in Figure 4.71.

Mezzanines are often used by modular producers to create elevated production and storage areas. If sufficient clearance is provided, major subassemblies can be produced on a mezzanine and even on the floor below a mezzanine (Figure 4.63). For example, in the overall factory design shown in Figure 4.72, a mezzanine might be useful to build subassemblies for roofs. These subassemblies are installed on the roof while it is being built on a roof jig and are installed on modules in Workstations 10 and 11. The same mezzanine might also store insulation used in the roof. If more floor space is needed, the entire roof building activity might also be placed on the mezzanine. The mezzanine shown in Figure 4.73 is used to build subassemblies and store materials. The floor space underneath the mezzanine is used to store insulation and build subassemblies for framing walls. The mezzanine shown in Figure 4.44 houses a framing table (in the background) and queue for side and marriage walls. The floor space underneath the mezzanine is used for a rough plumbing workshop.

The clear height of the facility is the distance between the floor and the lowest overhead obstruction: structural roof elements, lighting, sprinklers, HVAC and other obstructions). The clear height depends on many factors including:
- Module transport system – height of the module floor above the factory floor due to floor-mounted track, casters, wheels or other transport mechanisms
- Product design – floor depth, wall height, ceiling depth, roof height (module width, roof pitch), folding roof design
- Production process – for example, those activities that require the roof to be partially or fully unfolded (installing electrical, plumbing, insulation, decking, shingles, fascia/soffit, wrap)

- Mezzanines and any production activities, material storage or other uses intended for the mezzanine
- Cranes – as discussed in Section 4.6.6.1 above
- Future facility flexibility

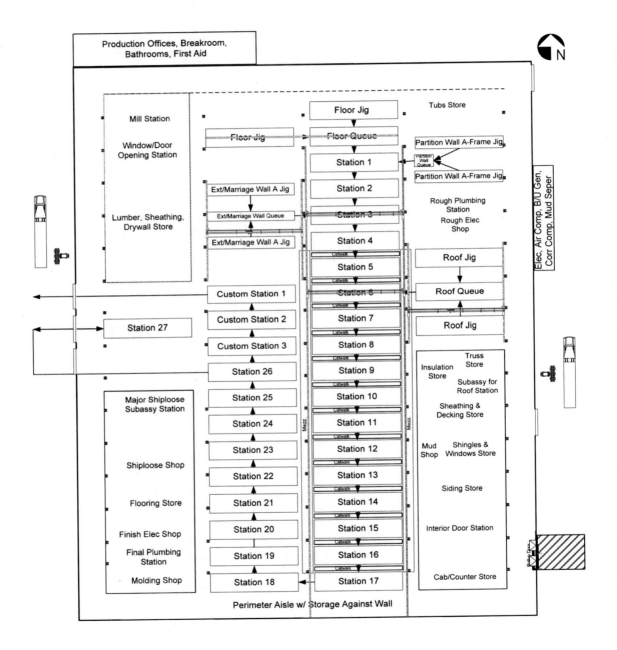

Figure 4.72 Overall factory design with J-shaped sidesaddle line layout for high production volume

Figure 4.73
Mezzanine used
for light
production and
material storage
(note lumber
and insulation
storage
underneath)

Many producers perform production activities and store materials outside the factory facility. They do this to accommodate unexpected growth in demand or customization in their current facility, to use an available smaller facility or to reduce the floor space and cost of a new facility. Outside activities may be performed on an extension of the production line (Figure 4.74) or anywhere in the yard or in an out-building after loading onto a carrier (Figure 4.75). The module must be weather tight before moving outside: roof wrapped, walls sheathed and windows installed. Material storage may be provided in a separate storage shed (Figure 4.76) or against the factory wall under a roof overhang or a freestanding shed (Figure 4.77). Performing an activity outside risks weather-related damage to the module. It also leads to lower productivity due to weather and limited access to supporting workshops, materials and supervision. Storing materials outside the factory can result in material damage, delayed material delivery to the line, lower productivity in material handling and tracking mud into the factory.

Figure 4.74
Shotgun production line
extending outside
factory facility

Figure 4.75
Workstations in yard used
for cleanup after modules
are loaded onto carriers

Figure 4.76
Separate shed used for
material storage

Figure 4.77
Freestanding shed against
factory wall used for
material storage

4.8 DESIGN THE WORKPLACE

The final phase of factory design is to add detail to the design of each workplace. Primary materials and equipment are arranged in each workstation, workshop, stockroom, warehouse and other production support area. The lean 5S visual management system introduced in Section 2.2.2.1 offers sound guidance for arranging the workplace. A foundational 5S concept asserts that only those items used in the workplace should be located in the workplace. A second 5S concept declares that frequently-used items should be located so that they are easy to find, access and return. A third 5S concept states that materials and equipment should be located to facilitate cleaning, inspection and maintenance. The following guidelines to reduce wasted motion in the workplace serve as working corollaries:

- Locate items by frequency of use, with commonly used items within reach (close and in front of the worker).
- Develop short, straight travel paths for workers and materials that do not cross one another.
- Use labels, signs and outlines to visually identify each item, where it should be located and how much should be there.
- Make storage locations larger than necessary to facilitate access.
- Maintain comfortable posture for workers with comfortable motions.
- Use mechanical assists when necessary to eliminate worker fatigue and injury.
- Kit large, non-standard materials (trusses, tubs/showers, cabinets) for each module and deliver to the workstation when needed. Locate the primary inventory of these materials outside the workstation.

These workplace guidelines have broader analogies that were incorporated into larger scale design decisions earlier in the design process:

- The queue for a major subassembly should be located adjacent to the workstation on the line where it is set on the module. It should also be located adjacent to the workstation where it is built.
- The orientation of a major subassembly should remain the same through build, queueing and set.
- A workshop that supports an activity should be located adjacent to the workstation where that activity is performed.
- If there is not room to store all the material in the workstation where it is used, the material should be stored nearby.

A spaghetti chart such as that shown in Figure 4.78 is used to visually identify wasted motion and congestion in the workplace.

Using the 5S concepts, a detailed layout is developed for each workplace. Each layout documents the location of the following items:

- Production equipment – fixed equipment including framing jigs, table saws, chop saws, panel saws, gang nailers, adhesive foam spray systems and paint spray systems.
- Material handling equipment – fixed equipment including cranes, conveyors and storage rack.

- Subassemblies and materials – staging for larger items including modules, subassemblies, trusses, framing lumber, sheathing, drywall, insulation, doors, windows and cabinets.
- Aisles – used to access equipment and materials.

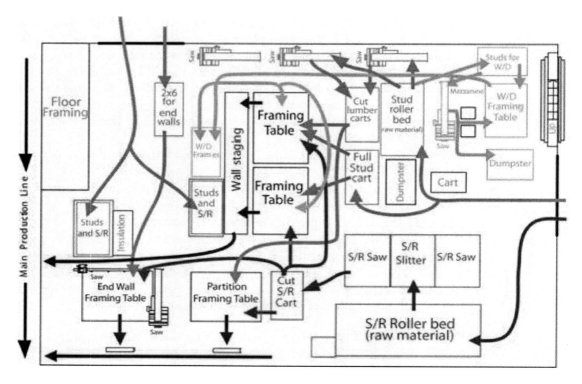

Figure 4.78 Example of a spaghetti chart for building partition walls [6]

The layout should also facilitate equipment maintenance and address housekeeping/environmental issues such as sawdust, drywall dust, spray foam adhesive, spray paint and drywall mud.

4.8.1 Case study

A recent study [6] demonstrates the use of lean concepts to redesign the wall build area for a modular producer. The wall build area is shown in Figure 4.79 and diagrammed in Figure 4.80. Sidewalls are built on one long framing table. Marriage and partition walls are built on a second long framing table. Walls are moved to the mobile staging rack using a bridge crane. The bridge is oriented parallel to the tables to facilitate handling. Short walls are staged on carts, and long walls are staged in the mobile rack. When short walls are needed on the line, the carts are pushed to the line. Long walls are moved to the line by pushing the mobile staging rack out from under the bridge crane and under one of the two monorail cranes that are used to move the walls to the line (Figure 4.81). Longer materials (framing lumber, drywall) are located on the long sides of the tables. Lumber is staged on blocking on the floor, and drywall is staged on stands for improved access. Materials are hand carried from staging to the tables. The working aisles between the long tables and the materials are not wide enough to replenish materials using a forklift truck. Instead, materials staged on the outside of the tables are replenished from the back using a forklift truck. Materials staged between the tables are replenished using the bridge crane equipped with a fork attachment. Scrap bins and cabinets for smaller materials and supplies are also located along the long sides of the tables. Pallets of smaller pre-cut components are staged in pallet rack along the perimeter of the mezzanine. These components are replenished by forklift truck. Headers are pre-cut under the mezzanine, moved by cart and hand-stacked on-end on the floor along the wall on the left side of the tables. Lumber used for the top and bottom plates is staged outside the factory because of its length. It is retrieved manually, spliced using the gang nailer and hand-carried to the tables. A chop saw is located near the end of the tables.

Figure 4.79
Wall build area
for modular
producer

End and gable walls are built on the short table near the bottom of the layout. When complete, they are staged on carts and moved to the line when needed. Framing materials are located in the two E-racks (cantilever racks) adjacent to the table. Materials are hand carried from staging to the table. Materials are replenished to the racks using a forklift truck. A chop saw is located near the table.

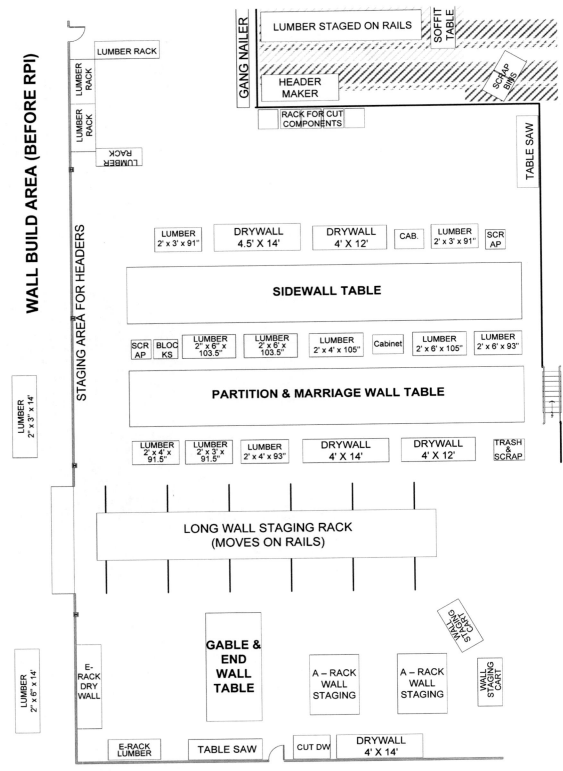

Figure 4.80 Diagram of wall build area for modular producer

Figure 4.81
Mobile wall
staging rack
used to move
walls under
bridge and
monorail cranes

The overall layout is relatively space efficient. There is little wasted space. Flow of heavy completed walls is mechanized, straight and efficient. Material is staged near the point-of-use, and staging locations are well-identified. Although materials are hand carried to their point-of-use, the flows are generally short and straight. Materials can be replenished easily with conventional equipment. Several opportunities related to the layout were identified. Rejected studs are stacked to the side of the tables, taking up valuable space. Lumber used to build the top and bottom plates is located outside the building, requiring lengthy travel carrying long (14') heavy material. In the winter, the lumber is covered with ice. The location of the gang nailer created interference with overall plant traffic and material flow. After joining, the plates are staged in the aisle near a break room, creating a hazard. Two workers are required to carry the long (40'+) joined plate from the gang nailer to the table. The plate must be reoriented 90 degrees during this move. The door near the small framing table is the main access-way between the offices and the factory. Employees using the door frequently interrupt work on the table.

To assist in identifying options, a template of the area was constructed, with movable cut-outs representing equipment and materials that could be relocated (Figure 4.82). Recommended improvements are shown in Figure 4.83. They include:

- Follow-up with vendors to reduce the number of rejected studs. Reduce the size of the area used to store rejected studs.
- Stack the same staged material, when possible, to free additional staging space.
- Move lumber used for plates inside to staging locations alongside the long tables.
- Install the gang nailer on a rolling cart that can be moved alongside the long table where the joined plate will be used.

- Reverse the hinges on the door so that foot traffic is guided away from the small table.

Figure 4.82
Template used
to identify
layout options

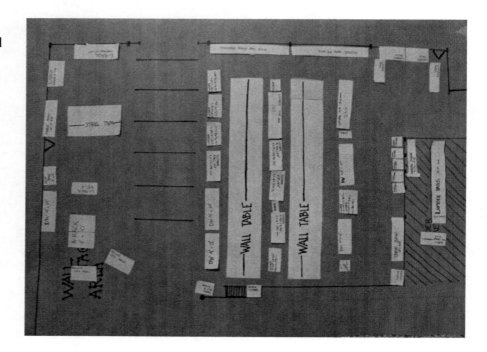

The study of the wall build area identified other critical issues that were not directly related to the layout. These included:

- Communication between factory management, area supervision and area workers.
- Sequencing production based on line needs.
- Kanban system of replenishment for cut components and headers.

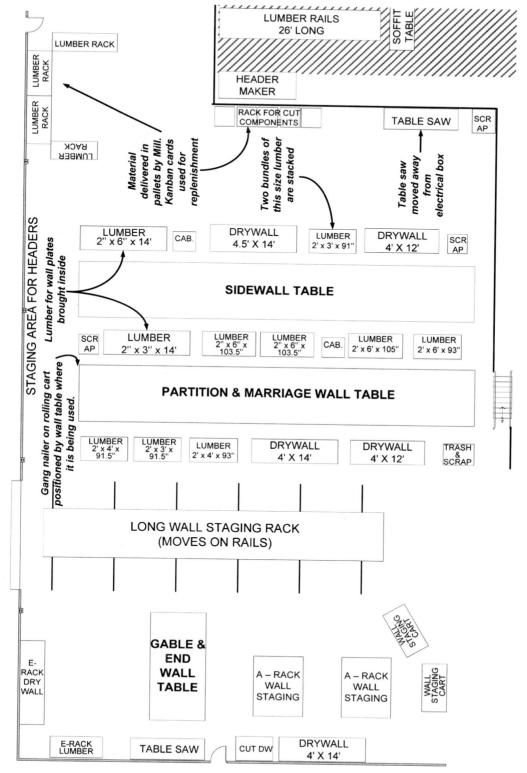

Figure 4.83 Diagram showing proposed improvements for wall build area

4.8.2 Handling building materials in the workplace

Building materials are critical to production performance in the workplace. It is far too common to see lengthy delays at vital operations such as wall and roof build because: materials are not there; the wrong materials are there; materials cannot be found; or materials had quality problems. More lengthy delays invariably disrupt flow on the line and propagate throughout the factory. Less obvious material-related delays are even more common: materials are staged far from their point-of-use; materials are staged where they are inaccessible (high in a rack or on the bottom of a stack); or materials are staged in the working aisles, creating congestion. Although shorter in duration, these delays reduce production efficiency, add to cycle time variation and contribute to flow disruptions throughout the factory. Therefore, handling building materials plays a prominent role in the detailed arrangement of the workplace. Workplace handling includes the replenishment, staging and use of materials. The discussion focuses on handling long building materials in the subassembly areas, since these scenarios are the most problematic. This includes handling framing lumber, sheathing, drywall and trusses in the floor, wall and roof framing workstations. Analogous concepts apply to other workstations that handle similar building materials.

Long materials are received on a flatbed trailer in unit loads called bundles or bunks (lumber only). Bundles are usually not palletized and are separated by blocking to allow handling by forklift trucks. When possible, materials should be handled in the same bundle in which it was received. This minimizes handling effort. Unpalletized bundles are moved transversely, using a vehicle equipped with forks. A counterbalanced forklift truck (Figure 4.37) can be used to move loads anywhere on the factory floor. It is well-suited for longer moves and moves requiring a lift. A pallet jack is better-suited for shorter moves at floor level or on a mezzanine where a lighter vehicle is required (Figure 4.84). Moving a bundle transversely requires a wide factory door and access aisles. Major access aisles should be at least as wide as the longest bundle plus necessary clearance. An even wider aisle is required if loaded vehicles are to pass each other. Aisle widths can be reduced by carrying a bundle lengthwise. This can be done using a narrow aisle sideloading forklift or palletizing the bundle and using a heavily counterbalanced forklift.

Figure 4.84
Pallet jack used for moving bunk
of studs on mezzanine

In the workplace, the objective is to minimize the movement required to retrieve materials for use, while conserving floor space. Staging material alongside the working aisles on the long sides of a framing jig serves to position material closest to the point-of-use. Staging multiple bundles of a high use material at several locations along the length of the framing jig can further reduce wasteful movement. However, it consumes valuable floor space which could be used by other materials. To conserve floor space, common staging locations can be located between two framing jigs, with each staging location accessible from both sides. This technique is used in Figure 4.83 between the two long framing tables. This technique is particularly useful when common materials are used on both framing jigs (for example, when a framing activity is replicated).

Staged materials can be oriented lengthwise (parallel to the working aisle) or widthwise (perpendicular to the working aisle). A widthwise orientation allows more bundles to be staged along the aisle, but creates several challenges including: wider workstation area, wider replenishment aisles, loss of staging utilization due to differing material lengths, and the need to handle bundles on pallets. Lengthwise staging orientation is assumed for the remainder of this discussion.

Long materials are usually staged on blocking on the factory floor (Figure 4.79). Short stands can be used to elevate staged material to a more convenient working height (Figure 4.79). Material can also be staged on the extended arms of a cantilever rack, allowing multi-level staging in the same footprint (Figure 4.85). Rack height is limited by several factors. Material must be easily accessible to the worker. When two levels of a rack are used for staging, it can be more difficult to reach materials near the back of the bundle on both levels. If a crane is used to move subassemblies or materials over the staging area, rack height must be limited to allow clearance for crane operations overhead. Rack height also limits load overhang when replenishing a long bundle on a narrow replenishment aisle. This is described in additional detail in the following paragraph. If overhead clearance is not limited, upper rack levels can be used for secondary storage of materials used in the area. Many materials can also be stacked on each other, with blocking between bundles, to further increase storage density. Density can also be increased by stacking a material deep, one bundle behind the other.

Figure 4.85
Material staged in
cantilever racks
(background) for use
on wall framing table

Regardless of staging configuration, it is important to keep working aisles narrow to minimize travel time and floor space. If materials are replenished from the front (using the working aisle), aisle width must be increased for forklift access. Note that aisle width can be narrower than bundle length by allowing the bundle to be moved while elevated over nearby obstructions (framing jig, staged material). This requires a low profile framing jig (a table instead of an A-frame), low profile staging equipment (instead of high storage racks) and a forklift truck (instead of a pallet jack). When replenishing material using this approach, the working aisle must be sufficiently wide to allow the truck to turn far enough so that the bundle can be deposited in its staging location. A 90 degree turn is required to allow the bundle to be deposited completely into its staging location. A partial turn (requiring less floor space) may allow the load to be dropped into the aisle and then pushed into the staging location using the forklift truck. The latter approach can only be used when the bundle is staged directly on the factory floor.

Several other approaches may be used to reduce the width of the working aisle. Both are used in the case study discussed in Section 4.8.1 above. Material can be replenished from the rear using a dedicated replenishment aisle. Materials staged on the factory floor, on blocks or in conventional pallet rack are accessible from the rear. Materials staged on cantilever rack are not accessible from the rear. The dedicated replenishment aisle requires additional floor space, but this can be minimized by sharing the aisle – also using it to replenish materials staged for use on another framing jig. A second approach to reduce the width of the working aisle is to replenish materials using a bridge crane equipped with a fork attachment.

Mechanization can reduce handling effort and floor space needs in the workplace. Material can be staged on a roller conveyor on the factory floor running alongside the working aisle. Bundles are replenished from the end of the conveyor without entering the working aisle, allowing it to remain narrow. When a bundle is depleted, a replenishment bundle is loaded at the end of the conveyor and pushed until the desired staging location is reached or until all gaps on the conveyor are filled and materials can be moved no further. Note that if different materials are staged on the same conveyor, staging locations cannot be dedicated to each material. Instead, a bundle is moved as other bundles are depleted and new bundles are replenished. Material can be oriented on the conveyor lengthwise or widthwise. It is easiest to load the conveyor from the end using the widthwise orientation. The widthwise orientation also increases the number of bundles that can be staged on the conveyor along the working aisle. However, it generally requires blocking or palletization to enable transverse flow on the rollers. Figure 4.86 shows a bunk of lumber loaded widthwise on a floor-level conveyor with blocking to facilitate flow. The widthwise orientation also widens the staging area. However, the same conveyor can be used to stage materials used on two framing jigs, one located on each side of the conveyor. Bundles may also be oriented lengthwise on the conveyor. However, this creates challenges loading the conveyor. If loading from the end, palletization and a heavy counterbalance forklift truck is required. To load the conveyor from the side, the working aisle must be widened (defeating the original purpose) or the conveyor must be extended beyond the length of the framing jig for loading. Conveyorized (or floor) staging can also be provided on the end of a framing jig, with bundles oriented widthwise to minimize the length of the area. This approach can be useful for staging custom materials or kits that must be replenished for each module – for example, for staging custom roof trusses (Figure 4.87). A second level of conveyors might be added to increase staging capacity for any conveyorized approach. However, this is likely to make access more difficult.

Figure 4.86
Conveyor used to stage framing lumber

Figure 4.87
Conveyor used to stage custom trusses

Mechanization can also reduce the effort of carrying materials to a framing jig. A popular handling device is the material bridge, a transfer car spanning the width of the jig table (Figure 4.88). The bridge rides on rails mounted to the factory floor on both sides of the jig table. The bridge is used to deliver large, heavy materials along the length of a long jig table. The bridge can be designed to tilt, allowing materials to slide off the bundle instead of being lifted (Figure 4.89). A transfer car can also run alongside a jig (Figure 4.90). However, this can make it more difficult to access any materials staged on the far side of the transfer car rails. A material bridge or transfer car can be used to distribute one or more different materials to a framing jig. If only one type of material is distributed, the car can also be used as the staging location for the material. If more than one type of material is distributed using the car (for example, drywall and trusses on a roof jig), then it will require loading and unloading the different materials for each module. Note that this is less of a problem for custom materials that are bundled for each module (for example, custom roof trusses). A possible solution is to use a larger car capable of carrying multiple materials. A bridge crane equipped with a fork attachment can also be used to distribute heavy materials along the length of a workstation (Figures 4.91–4.93). An industrial manipulator, mounted on a crane and fitted with custom end-effectors for gripping, can be used to lift and move heavy, bulky materials (sheathing, drywall) from the staging location to the point-of-use.

Figure 4.88
Material bridge used to distribute drywall on roof table

Figure 4.89
Tilting material bridge (left) used to distribute material on framing table

Figure 4.90
Transfer car
used to carry
roofing
materials
alongside roof
table

Figure 4.91
Bridge crane
equipped with
fork attachment
used to
distribute
trusses on roof
table

Figure 4.92
Loading area for
bridge crane used
to distribute
shingles on roof

Figure 4.93
Bridge crane used
to replenish
drywall to material
bridge on roof
table

Standard pallets are convenient for handling unit loads of many medium-sized materials. Many materials are received on pallets, such as the shingles shown in Figure 4.92. Other materials can be handled on pallets after pre-fabrication, such as the pre-cut components shown in Figure 4.73. Palletized material can be handled by a wide range of vehicles equipped with forks. Palletized materials can be staged directly on the factory floor or on a conveyor. If greater storage density is required, palletized materials can be stacked high (if stackable), deep (one pallet behind the other) or staged in pallet rack.

Some materials are best staged on dedicated specialty equipment. Vinyl flooring and carpet is received and staged in long rolls. These rolls can be staged in cantilever rack

(Figure 4.94) or in a powered vertical rotary rack that can deliver each roll to floor level for dispensing. Exterior doors are staged standing up in a special fabricated frame (Figure 4.95). Windows are received and staged in special stackable frames (Figure 4.96). Pipe can be staged in the plumbing workshop on light duty cantilever rack or pegs (4.97). Interior trim can be staged in the trim workshop on light duty cantilever rack or specialized carts (Figure 4.98). Insulation, which is bulky but light, is received loose stacked in truck-load quantities. It may be handled and staged by hand stacking the individual rolls or palletizing (Figure 4.73) for more efficient handling over longer distances.

Figure 4.94
Cantilever rack (background) used to stage rolls of vinyl flooring

Figure 4.95
Special fabricated frame used to stage exterior doors

Figure 4.96
Special stackable frames
used to stage windows

Figure 4.97
Light duty cantilever rack used to stage
pipe in plumbing workshop

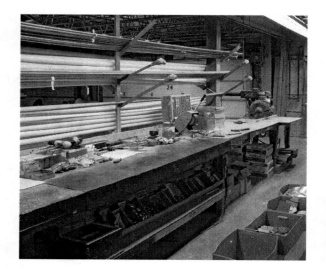

Figure 4.98
Specialized carts used to
stage interior trim materials
and pre-cut window trim
components

Components that are custom fabricated and kitted for a module in another workstation, workshop or stockroom can often be delivered to the workstation and staged on a cart (Figures 4.99–4.100). This simplifies handling, reduces the number of materials in the workplace, and provides greater mobility to teams serving multiple workstations on the line. When the tools and parts needed for an activity on the line are small, a cart can also be used as a mobile workshop (Figure 4.45).

Figure 4.99
Cart with cantilevered arms used to stage kit of pre-cut framing components

Figure 4.100
Cart used to stage kit of pre-cut trim components

Small parts are received and handled in their shipping cartons. They can also be staged in their shipping carton (Figure 4.97) or emptied into a bin (Figure 4.101). Because of the relatively small size, backup stock can often be stored in the workplace near the primary staging location.

Figure 4.101
Bins used to stage small parts in plumbing workshop

4.8.3 Secondary storage for building materials

Ideally, all materials are procured from dedicated vendors located near the factory. Ideally, these vendors offer value pricing with small order quantities, low shipping cost, short lead time, and high quality. As a result, materials can be ordered in small quantities on a just-in-time basis, received near their point-of-use and staged for immediate use. Practically, material procurement is far from this ideal. Vendors are located far from the factory. The factory, and indeed the whole modular industry, represents only a fraction of the vendors' business. As a result, vendor performance is less than ideal: prices are unstable with large volume discounts, lead times are uncertain and incoming quality is unreliable. This results in larger order quantities, larger safety stocks and higher inventory levels. Larger inventories of materials cannot be staged in the workstation where they are used without reducing efficiency and adding floor space. The situation is further complicated by other factors:
- Many materials are used in a standard home design.
- Design variations resulting from model offerings, options and customization multiply the number of materials that are used.
- Some materials are used at multiple workstations.

To reduce inventory levels and the resulting space requirements in the workplace, it is necessary to maintain secondary storage locations for certain materials. These materials are stored in their secondary storage location upon receipt and replenished to their staging location in the workplace only when they are needed. Ideally, replenishment is controlled using a kanban signal. Material can be stored inside or outside the factory. Outside storage may be provided under factory roof overhangs or in separate sheds and warehouses. Storage can be centralized or distributed throughout the factory. Centralizing storage allows greater control and storage density. Expensive, hard-to-get, or other critical materials can be centralized in a stockroom with controlled access. Tools can be centralized in a crib. Larger, high volume building materials can be centralized in high density storage areas where stackable materials are stacked high and deep (Figure 4.102) and non-stackable and/or lower volume materials are stored in high-rise racks (Figure 4.103). Centralized storage encourages the use of more specialized storage/retrieval equipment, such as a sideloading forklift trucks that are able to access high storage racks using a narrower aisle. It is also much easier to manage (reorder, cycle count, physical inventory) materials that are used at multiple workstations when they are centralized.

Figure 4.102
Block stacked
materials are stored
high and deep

Distributing storage allows materials to be located closer to their point-of-use. Philosophically, distributed storage is more consistent with lean production. It prevents institutionalizing the warehouse and all that it represents: large inventories, overhead and waste. Instead, distributed storage can be used to enforce small inventory levels and kanban-controlled reordering and replenishment.

Figure 4.103 Tall cantilever rack used to store non-stackable or lower volume materials

The factory designs shown in Figures 4.71 and 4.72 utilize both distributed and centralized storage. Various materials are distributed around the perimeter of the facility. Lumber, sheathing and drywall are centralized near the mill area.

4.9 REFINING THE DESIGN

During each phase of the design process, there are opportunities to reexamine earlier design decisions. This is particularly important when it is discovered that an earlier decision limits later design options. For example, did the line layout provide adequate floor space to design an efficient workplace? Did it result in wasted space? Hopefully, it is possible to capitalize on these opportunities by refining the earlier design to reflect the improved understanding of the problem. Now, after completing the first iteration of the design process, it is important to revisit the design as a whole to confirm that it satisfies the overall goals of the effort:

- Provide a safe and satisfying work environment.
- Produce a product that meets customer expectations for variety, performance, delivery time and quality.
- Reduce waste to cut cost, decrease price and increase profitability.

These goals are largely attained by accomplishing the following objectives:

- Provide static capacity – reduce average cycle time for each activity so that it is less than or equal to TAKT time.
- Accommodate variation – reduce cycle time variation. Reduce the impact of any remaining cycle time variation.
- Produce quickly – shorten overall production cycle time.
- Produce efficiently – consume no more resources than needed (floor space, equipment, materials, labor).

This is an opportunity to study the overall design, reflect on its many diverse elements and refine the design to better accomplish these goals and objectives. For example, the factory's ability to accommodate variation is driven by complex interactions between product design/architecture, basic production processes, activity structure and flow, configuration of the line, material handling system configuration and design of the workplace. Better integration of these elements often yields improved capability to accommodate variation, without detracting from other objectives.

4.10 REFERENCES

1. McCoy, A., W. Thabet and R. Badinelli, "Towards Establishing a Domain Specific Commercialization Model for Innovation in Residential Construction", *Construction Innovation: Information, Process, Management*, 8(2), 137–155, 2008.
2. Oglesby, C., Parker, H., and Howell, G. *Productivity Improvement in Construction*, McGraw-Hill, New York, 1989.
3. *Pre-cast Concrete Panelizing*, Housing Constructability Lab, http://www.housingconstructabilitylab.com/pages/precast.htm, October, 2009.
4. Broadway, R. and M. Mullens, "Shop Floor Information Systems for Industrialized Housing Production," Industrial Engineering Research '04 Conference Proceedings, Houston, May, 2004.
5. Mullens, M., *"Data Collection and Predictive Modeling in Industrialized Housing,"* Presented at IFORS 2005 Conference, Honolulu, July 2005, http://www.housingconstructabilitylab.com/pages/modular%20housing%20shop%20floor%20control.ppt, October, 2009.
6. Manufactured Housing Research Alliance, *Pilot Study: Applying Lean to Factory Homebuilding*, U.S. Department of Housing and Urban Development, Office of Policy Development and Research, Washington, D.C., July, 2007.
7. Nahmens, I. *Mass Customization Strategies and Their Relationship to Lean Production in the Homebuilding Industry. Ph.D. Dissertation*, University of Central Florida, Orlando, FL, August, 2007.
8. Mullens, M. and R. Toleti, "A Four Day Study Helps Home building Move Indoors," *Interfaces*, 26(4), 13-24, July, 1996.
9. Manufactured Housing Research Alliance, *Develop Innovations in Manufacturing Processes through Lean Production Methods*, U.S. Department of Housing and Urban Development, Affordable Housing Research and Technology Division, Washington, D.C., October, 2005.

CHAPTER 5
IMPLEMENTATION: REALIZING THE PROMISE

Chapter 5 addresses how the factory design developed in Chapter 4 is implemented. It is only through successful implementation that the full promise of modular homebuilding is realized. Implementation occurs in several stages, including planning, procurement, fit-out, startup and ongoing operation. Each of these stages is discussed below. This discussion is not meant to be comprehensive, but only to serve as an overview of the implantation process.

5.1 PLANNING

One of the first steps in the implementation process is to assemble the senior management team that is responsible for all planning. Key players include a general manager and his/her direct staff, who are responsible for sales/marketing, engineering/design, purchasing, manufacturing, human resources, quality and finance/accounting. Ideally, the senior management team is in place and functioning at the start of the factory design effort, allowing critical collaboration in all early planning decisions. Primary planning responsibilities include:

- Sales/marketing
  - o Sales plan – sales forecasts and pricing
  - o Marketing plan – market summary (market analysis, trends, issues, strengths/weaknesses versus competitors), marketing strategy (target market, sales message, critical issues), launch strategy, marketing mix (promotion and advertising, sales support including training and materials, promotion channels such as media/telemarketing/website)
  - o Marketing launch
  - o Recruiting and training builders
- Engineering/design
  - o Product portfolio – standard models, options, and customization allowed
  - o Bill of Materials – materials needed
- Purchasing
  - o Supply chain strategy for purchased materials – make versus buy decisions, inventory management (inventory levels and reorder discipline/quantities)
  - o Identification and qualification of suppliers
  - o Solicitation of quotes
  - o Negotiation of contracts
- Manufacturing
  - o Capacity plan
  - o Factory design
  - o Production staffing estimates
  - o Capital and manufacturing cost estimates

- Human resources
  - Compensation and benefits plan
  - Safety plan
  - Recruiting, interviewing, hiring and training staff
- Quality
  - Inspection plan – purchased materials, work-in-process, finished goods
  - Warranty service plan
  - Long-term continuous quality improvement plan – measurement of key performance metrics, development of continuous improvement teams, recognition/reward of accomplishments; note that continuous improvement efforts should be integrated with lean production efforts
- Finance/accounting
  - Plan for financing and operating the new factory – Raising initial capital, managing cash flow
  - Accounting system

5.2 PROCUREMENT

After the design of the new factory has been completed, it must be procured. Procurement involves locating, building and equipping the new factory. Factory location is determined while the factory is being designed. Factors that must be considered include: proximity and access to markets and materials, availability and cost of skilled labor, labor climate, availability and cost of industrial sites, industrial construction costs, taxes, political/regulatory climate and incentives offered by state and local government. Factors that must be considered when selecting a specific site include: zoning restrictions, access to transportation, utility access, site size and configuration, site work required and purchase/lease/financing terms.

A second key procurement decision is whether to build a new greenfield facility or utilize an existing facility. An existing facility may be new or previously used and subsequently shuttered. A new facility can meet the precise needs of the factory design. However, it is usually more costly and time consuming. An existing facility can often reduce capital costs and allow production to begin sooner. Procuring an existing facility and ramping up production can take 6–12 months, while a new facility can take twice that long. In an emerging market with few competitors, the ability to enter the market early can be vitally important to both short and long-term business success. However, due diligence is required before acquiring an existing facility. The facility must meet the locational needs discussed in the previous paragraph as well as the design needs identified during the factory design effort. The facility must either conform or be reconfigurable to these design needs. Most facilities fail to meet these basic requirements. Even if an existing facility provides the required floor space, its configuration is likely to be incompatible with the factory design: column spacing is too narrow; clear height is too low; the floor is at multiple levels; the shape does not accommodate a line; or the crane system is inadequate. Although incompatible features can be changed, it is always at the cost of both money and time, which erodes the advantage of the existing facility. An existing facility can also compromise production performance. Modular producers who do use existing plants have experienced the following disadvantages:

- Insufficient floor space – limits basic capacity; requires some work to be performed outside; limits queueing and ability to accommodate variation without disrupting flow.
- Insufficient clear height – limits home design; prevents the roof from being extended after set on the module.
- Inadequate crane system – limits capacity and flexibility.

In general, the more compatible an existing facility is with the factory design, the more likely it is to provide the expected competitive advantage.

If a new greenfield facility is to be built, the procurement process requires two phases: detailed design and construction. In the first phase, architects/engineers use the factory design to develop the specifications, including detailed drawings, needed to construct the facility. In the second phase, a construction firm builds the facility to meet the specifications prepared in the first phase. Using the traditional design-bid-build approach to construction project delivery, each phase is contracted independently. The second phase is contracted after the first phase contract is completed. Each phase requires the following activities: develop a request for proposal (RFP); identify and qualify vendors; review vendor proposals; negotiate the contract; and manage the work. The RFP for each phase includes the primary product of the previous phase together with the legal terms and conditions of the contract. For example, the RFP for the first phase includes the factory design, and the RFP for the second phase includes the specifications needed to construct the facility.

The procurement process can be expedited using a design-build approach, in which a single contract is awarded for both design and construction. The contractor may be a large, integrated firm with both design and construction capabilities in-house, an architectural/engineering firm that subcontracts construction, or a construction company that subcontracts design. The design-build approach can save months in the procurement cycle by eliminating the second contracting process and streamlining the remaining value-adding activities. Streamlining opportunities include:

- Preparing only one set of specifications and detailed drawings, which represent the facility as it will actually be built.
- Parallel tasking – obtaining permits and beginning construction while the design is being finalized.

The design-build approach has other advantages for the modular producer (owner):

- More accurate construction cost estimates are available sooner. Architects/engineers that design the facility work directly with construction staff that will build the facility. Thus, design staff have a better understanding of the construction technologies and materials that will be used to build the facility. This reduces the risk of inaccurate (usually low) preliminary construction cost estimates and subsequent "finger-pointing" between design and construction. More accurate cost estimates also allow better tradeoffs earlier in the design process when they have maximum value.

- Communication is simplified. The owner has a single point of contact with the contractor. This contact serves as an information hub, routing information throughout the contractor organization, to subcontractors, to third-parties and to the owner.
- Accountability is clear. The contractor is responsible for designing and building the specified facility, within the budget and schedule. "Finger-pointing" between design and construction is eliminated, since the same contractor is responsible for both. This reduces the owner's risk if problems arise.

These advantages are maximized when the single-source contractor has both design and construction capabilities in-house.

The design-build approach has one major disadvantage – it purposely removes the independence between design and construction and thus the checks and balances inherent in the design-bid-build approach. Contractor bids are likely to be higher, since the facility is not fully specified before contracting. Higher project costs must be balanced against financing savings from reducing construction loan duration and additional revenue generated by earlier occupancy. After the contract is awarded, the design focus on adding value for the owner can also be distorted by the contractor's objective of delivering the project at the lowest cost.

Construction may also be expedited by using pre-engineered steel building systems. These systems use highly-engineered prefabricated components to build the facility. Like modular homebuilding, prefabrication of the factory components allows on-site construction to be performed faster and more efficiently. Since site work can be performed while the pre-fabricated components are being manufactured, the overall construction project cycle can be reduced.

Production equipment is procured while the facility is being designed and built. Overhead crane systems are usually procured as part of the facility, since it is often more efficient for cranes to utilize the facility structure. Standard equipment is purchased from conventional suppliers. Custom equipment is purchased from custom fabricators or supplied internally using factory staff. Equipment procurement involves: identifying equipment vendors, their products and pricing; developing equipment specifications, including detailed design drawings for custom equipment; soliciting quotes; negotiating contracts; and installing the equipment. To expedite startup, equipment is manufactured while the facility is being built so that it is ready for delivery as soon as the facility is occupied. It may also be possible to procure used equipment at considerable savings in cost and lead time.

5.3 FIT-OUT

Fit-out is the brief period between building occupancy and startup of production operations. During this time, equipment must be delivered, installed and checked-out to assure that it is ready for startup. To minimize the duration of the fit-out period, equipment procurement must be synchronized with the facility construction schedule. This allows delivery and installation to begin as soon as the facility is occupied.

5.4 STARTUP

Startup is the period between the start of production operations and reaching full capacity. During the startup period, production is usually carefully controlled, starting at a very low production rate and gradually ramping-up to full capacity. Startup provides a break-in period for the new production system, allowing on-the-job training and identification of problems in specific equipment and the overall production system. At the lower production rate, a problem can be resolved with minimal impact on the rest of the production system or on the customer. Although staffing levels during startup may be constrained to match production rates, they are likely to be greater than actually needed to account for training and the anticipated near-term growth. Ramp-up of production volume is gradual – for example, adding one module to the daily production schedule each succeeding month.

5.5 ONGOING OPERATION

It is in this final stage, after successful factory implementation, that the full promise of modular homebuilding is realized. However, it is important to emphasize that the creation of a new, high performance factory does not guarantee production excellence. History has repeatedly shown that an organization must continue to innovate. The production strategy developed in Chapter 2 provides a roadmap for continuing innovation. It focuses on five critical areas: supply chain management, capacity management, lean production, quality management and mass customization. Continuing innovation must occur at every level of the organization, starting in the workplace. Workers must be empowered to continuously improve their workplace and the processes that they perform. Worker involvement can be organized through kaizen (continuous improvement) teams, as discussed in Chapter 2.

Continuing innovation must also address broader, cross-cutting issues. New home designs, materials, equipment and systems must be developed that provide strategic advantages. Innovations must benefit multiple stakeholders – the producer, homebuilder, homebuyer, society and the environment – making adoption a win-win proposition. For example, a truly successful innovation might reduce physical exertion on the worker, reduce defects, smooth production flow, reduce production cost, reduce set and finish effort in the field, improve ongoing home performance for the homebuyer, and leave a smaller carbon footprint for the environment. This chapter introduces three innovative technologies with the potential for cross-cutting benefits. They include Optimum Value Engineering (OVE) for framing, supersize building materials and real-time shop floor information systems. Although these technologies are only a small sample of potential innovations, they serve to demonstrate how innovative technologies can add value to factory homebuilding.

5.5.1 Optimum Value Engineering for Framing

5.5.1.1 What Is It?

Optimum Value Engineering (OVE) for framing, also called advanced framing, is a set of framing design practices that limit framing components to those needed to support the structure [1,2,3]. Some OVE practices include:

- Two stud corners – instead of 3 stud corners.
- 2"x6" studs on 24" centers aligned with roof trusses – instead of 2"x4" or 2"x6" studs on 16" centers.
- Single top plates – instead of double top plates.
- Single headers – instead of double headers.

5.5.1.2 What Are Its Advantages?

OVE framing reduces the amount of wood required to build a home. Savings estimates (board feet) range from 5-10% [2] to 19% [1]. This reduces lumber cost and environmental impact accordingly. OVE framing requires 30% fewer sticks/pieces of framing lumber [2], speeding the framing process and reducing framing labor. Labor cost savings are estimated to be 3-5% [3]. Cost savings should increase profit for the producer/homebuilder and/or reduce price for the homebuyer.

OVE framing also reduces energy consumption for heating and cooling the home. The deeper wall cavity allows more insulation (when compared to 2"x4" framing), and the reduction in energy conducting studs can save up to five percent of the energy required annually for heating and cooling [3]. This results in lower utility bills for the homebuyer and conservation of scarce natural energy resources.

5.5.1.3 What Are the Obstacles to Adoption?

OVE framing has not been widely adopted by conventional site builders or modular producers. Site builders argue that any change is difficult [1,2]. Drawings must be revised. Builder staff, subcontractors and code officials must be educated. Framing subcontractors have not been willing to reduce prices, arguing that the process is more complicated and does not save labor. Builders are concerned that cost savings may not offset the perception that they are cutting corners.

Using standardized processes, specialized equipment and dedicated framing teams, modular producers are more likely to capture potential OVE cost savings. However, modular producers are already perceived by some homebuyers as offering an inferior product and must take care to insure that OVE framing does not reinforce negative homebuyer perceptions. Modular producers must also demonstrate that OVE framing can be integrated with existing processes – such as lifting an OVE wall (with a single top plate) in the factory and lifting a completed module on the construction site.

5.5.2 Supersize Building Materials

5.5.2.1 What Is It?

Supersize building materials are larger versions of conventional building materials. For example, a vendor developed supersize drywall 8' wide and up to 24' long for the modular industry.

5.5.2.2 What Are Its Advantages?

Supersize building materials reduce the number of pieces that must be handled, assembled and finished. For example, supersize drywall has been used for interior ceiling and wall applications, reducing the number of sheets and minimizing seams. Drywall screws were replaced with spray foam adhesive, eliminating holes. Reducing seams and screw holes reduced drywall finishing work, a leading cause of rework and bottlenecks. Manual handling effort was minimized by equipping workers with special industrial manipulators and mechanized material bridges specifically designed for handling the supersize drywall.

5.5.2.3 What Are the Obstacles to Adoption?

Supersize building materials cannot be used effectively on the construction site because of their size and difficulty in handling manually. However, a factory can be properly equipped to easily handle the materials. A risk to early adopters of supersize materials is the commitment of suppliers. For example, when the price of the supersize drywall rose substantially after introduction, expected demand never materialized. The supplier pulled the material from the market, leaving early adopters with sizable investments that could not be recovered [4].

5.5.3 Real-time Shop Floor Information System

5.5.3.1 What Is It?

A real time shop floor information system provides the information infrastructure to guide sound production decision-making. An example is the Status Tracking and Control System (STACS) [6]. STACS was developed as a working prototype to demonstrate the concept of real time labor data collection and reporting in the modular factory. Production workers used wireless laser scanners to report their current work assignments. Scanned information was transmitted to a central database server where it was stored and used for reporting. STACS reporting was web-based and provided both real time manufacturing status and summaries of historical production performance. Real time production performance could be monitored from the web-based STACS Dashboard. "Clicking" on any item on the Dashboard displayed corresponding real-time details. Historical results were used for a variety of analytical and management purposes including: 1) the development of analytical labor estimating models (these models can be used to estimate labor requirements for product costing, production scheduling and labor planning) and 2) as a baseline for continuous improvement efforts.

5.5.3.2 What Are Its Advantages?

When given a daily/weekly production schedule, a real time shop floor information system can use historical labor data that it has collected to predict production cycle times and forecast potential flow disruptions on the line. Line supervision can use the forecast to develop a response plan to minimize the disruption. A real time shop floor information system can also identify emerging problems (bottlenecks, line vacancies, out-of-station activities) and signal the need for immediate action.

5.5.3.3 What Are the Obstacles to Adoption?

An early prototype of STACS was tested in several modular factories. Test results demonstrated that production workers could operate the system and capture useful information. However, their reporting was not consistent enough for the system to produce reliable results [7]. In general, the modular factory and its workforce are not tightly controlled. Thus, any information system requiring real-time worker input would require a change of culture.

5.6 REFERENCES

1. Binsacca, R., "A Fresh Look: A Focus on High-performance, Energy-efficient Housing May Finally Push Advanced Engineering and Framing Techniques into Common Practice, *Builder*, http://www.builderonline.com/high-performance-building/a-fresh.aspx?printerfriendly=true, 4/6/2010.
2. Lstiburek, J., "Building Sciences: Advanced Framing", *ASHRAE Journal*, 51(11), November 2009.
3. NAHB Research Center, Southface Energy Institute, U.S. Department of Energy's Oak Ridge National Laboratory, and U.S. Department of Energy's National Renewable Energy Laboratory, *Technology Fact Sheet: Advanced Wall Framing - Build Efficiently, Use Less Material, and Save Energy!*", Office of Building Technology, State and Community Programs, Energy Efficiency and Renewable Energy, U.S. Department of Energy, Washington, D.C., http://www.energystar.gov/ia/home_improvement/home_solutions/doeframing.pdf , 5/1/2010.
4. Mullens, M., "Production Flow and Shop Floor Control: Structuring the Modular Factory for Custom Homebuilding" *Proceedings of the NSF Housing Research Agenda Workshop*, Feb. 12-14, 2004, Orlando, FL. Eds. Syal, M., Mullens, M., and Hastak, M. Vol 2, pp. Focus Group 1.
5. Hopp, W. and Spearmann, M., *Factory Physics: Foundations of Manufacturing Management*, 2nd Edition, Irwin/McGraw Hill, NY, 2001.

6. Broadway, R. and M. Mullens, "Shop Floor Information Systems for Industrialized Housing Production," *Industrial Engineering Research '04 Conference Proceedings*, Houston, May, 2004.
7. Mullens, M., "Data Collection and Predictive Modeling in Industrialized Housing," Presented at IFORS 2005 Conference, Honolulu, July 2005.

INDEX

A

Activity, 98–104
    adding labor, 130–131
    batching, 142
    decomposing, 132, 141
    estimating labor requirements,
        99–103, 130–131
    mobility, 42, 132–136
    replicating, 131
    shifting (in time), 142
    unconventional, 139–140, 141
Air casters, 176
Aisle, 195, 208, 209, 210
Asynchronous flow, 132–136

B

Batch line flow, 144
Bathroom, 193
Bottleneck, 20–21, 31, 35–36, 39,
    132–136, 137
Breakroom, 193
Bridge crane. *See* Crane
Building codes. *See* Regulatory
    requirements
Build-in-place, 142–143, 165

C

Cabinets
    install, 82-83, 103, 137–138
Capacity, 19–20, 110–111
    management, 30– 32
    plan, 97–98
    utilization, 19
    vs. plant size, 194
    vs. workstations, 145
Cart, 136, 177, 215–217
Casters, 173
Catwalk. *See* Roof access
Ceiling. *See* Roof

Clear height, 191–193, 197
Column, 149, 19 –192
Cost. *See also* Labor cost; Material
    cost; Overhead cost; Service
    cost
    modular vs. site-built, 11–12.
Continuous flow, 34–37, 42
Continuous improvement, 23, 37–38
Countertops, 83
Conveyor, 174, 210
Crane
    building material, 208–214
    module, 177
    subassemblies, 177–193
Critical path method, 38
Cross-training, 8, 42, 136
Customer satisfaction, 14, 16, 22
Cut framing components, 48–49
Cycle time, 31
    estimation, 98–103
    modular vs. site-built, 11, 23
    variation, 6, 35–37, 39–43, 102–
        103, 132–133

D

Design-build contracting for the
    new factory (vs. design-bid-
    build), 225–226
Dock (loading), 195
Door
    exterior. *See* Window
    factory, 195
    interior (build interior door
        subassemblies), 84–85
    interior (install interior doors),
        85
Drywall
    hang, 81
    sand and paint, 82
    tape and mud, 19, 20, 21, 81,
        102, 104, 139, 142
    supersize, 230

E

Electric. *See* Rough electric; Finish
  electric
End walls
  build, 56, 126
  set, 59
Engineering. *See* Product design

F

Factory design
  goals and objectives, 95, 110
  management team, 223
  overall layout, 193–200
  process, 95–222
  refinement, 220–221
Factory within a factory, 43, 138
Fascia and soffit installation, 76
Finish electric installation, 84
Finish plumbing
  build subassemblies, 83
  install, 84
Finish process (on-site), 4, 23
Fit-out, 226–227
Five S. *See* Visual management
Flexible manufacturing system, 43,
  128
Floor
  build, 50, 143–144
  material handling. *See* Material
    handling (subassemblies)
Flooring installation, 86
Framing jig
  A-frame, 52
  table, 52

G

Greenfield (new construction) vs.
  existing facility, 224–225

H

HUD Code, 4

I

Implementation, 223–231
Incentives, 136
Information systems, 42, 136, 230–
  231
Insulation
  roof, 71
  walls, 77–78

J

JD Powers, 14, 16, 22, 23
Just-in-time, 34–37

K

Kaizen, 37
Kitting, 126–127

L

Labor
  absenteeism, 22
  cost, 12, 18
  factory vs. site, 8
  flexibility, 136
  involvement, 37–38, 227
  turnover, 22
Lean construction, 38–39
Lean production, 19, 29, 30, 32–38,
  42, 127–128, 201–207
Line. *See* Production line
Line balancing, 99
Load module on carrier, 89–90
Location (of factory), 224

M

Management, 9, 223
Market share, 14–15
Marketing issues, 17

Marriage wall
    build, 57, 202–207
    material handling. *See* Material
        handling (subassemblies)
    set, 59
Mass customization, 30, 39–43
Material
    bridge, 211–212
    cost, 12, 29
    management, 7, 13
    supersize, 230
    waste, 7, 13
Material handling
    building material, 193–200,
        208–220
    module, 173–177
    subassemblies (floor, walls,
        roof), 177–193
Mezzanine. *See* Roof access
Mill. *See* Cut framing components
Miscellaneous finish items, 86
Modular home, 1
Modular homebuilding, 1–6
    advantages, 6–14
    challenges, 15–24
    vision, 25
Module, 1, 45
    handling. *See* Material handling
        (module)
    number per home, 45
    size, 45–48
Molding, 85
Monorail crane. *See* Crane

N

National Housing Quality Award,
    23

O

Office, 193, 194
Optimum Value Engineering
    (framing), 228
Outsourcing, 41, 127, 142
Overhead cost, 12, 19, 29

Overtime, 21, 31, 98, 132–136

P

Palm Harbor Homes, 23
Panelized homebuilding, 4
Parallel production, 6, 131, 140–141
Partition wall
    build, 52, 202–207
    set, 57
Path
    critical, 111–118
    longest, 112–124
    longest sub-path, 118–124
Planning (overall business), 223–
    224
Plumbing. *See* Rough plumbing;
    Finish plumbing; Plumbing in
    floor
Plumbing in floor, 88
Postponement, 41
Precedence relationships, 104, 111–
    124, 135, 142
Pre-engineered steel building
    system (for new factory), 226
Prefabrication, 1–6
Process
    activity. *See* Activity
    flexibility, 41–43. S*ee also*
        Activity mobility; Crane;
        Labor flexibility
    improvement, 127–128. *See also*
        Lean production
    technology, 6–7, 17–18, 35, 52,
        98–99, 127–129
Procurement (factory), 224–226
    factory equipment, 226
    factory facility, 224–226

Product
    architecture, 40–41, 96 –97,
        124–127
    choice, 1, 19, 20, 30, 39–43, 45,
        96
    customization. *See* Product
        choice
    design, 9, 16, 45, 96–97
    movement. *See* Asynchronous
        flow; Synchronous flow;
        Batch movement; Material
        handling
Production
    activity. *See* Activity
    builders, 23
    leveling, 34
    rate, 19, 30–32, 97–98, 99, 146,
        227
    strategy, 29–39
Production line, 6, 132–136. S*ee
    also* Sidesaddle line layout;
        Shotgun line layout; Build-in-
        place
    design, 144–193
Proximity rule, 35
Pull replenishment system, 36, 137
Pulte Homes, 23

Q

Quality, 9–10, 14, 19, 22–23, 30,
    36, 224, 227
Queue, 35–37, 104, 137–138

R

Rack. *See* Storage rack
Rails (tracks), 174
Regulatory requirements, 1, 17
Rework, 21, 23
Roof. *See also* Insulate roof; Sheath
    roof; Shingle roof; Fascia and
    soffit installation; Wrap roof
    for shipment; Material
    handling (subassemblies)
    access to, 148, 167–173

build, 65
build subassemblies for, 63–64,
    140
hinged, 45
set, 67, 137
Rough electric
    install in walls, 59, 141
    install in roof, 69
Rough plumbing
    build subassemblies, 61
    install in roof, 70
    install in walls and tubs, 62

S

Safety, 12–13, 172
Sales plan, 97
Service cost, 23
Setup time, 31, 34, 98, 142
Set process (on-site), 4
Sheath
    roof, 72–73, 126
    walls, 79
Shift schedule, 21, 31, 97–98
Shingle roof, 74–75, 99, 126
Shiploose
    build major shiploose
        subassemblies, 92
    load material, 87
Shotgun layout, 155 - 164
Side wall
    build, 54, 126, 202–207
    set, 59
Sidesaddle line layout, 146–154
Siding and trim installation, 80
Single piece flow, 34
Site-built homebuilding, 4
Slack time, 38, 142
Standardization, 33–34
Startup, 227
Stockroom, 193, 194, 201, 217, 219
Storage. *See* Material handling

Storage rack
  cantilever. *See* Material
    handling (building material)
  elevated. *See* Material handling
    (subassemblies)
  mobile. *See* Material handling
    (subassemblies)
  pallet. *See* Material handling
    (building material)
Subassembly. *See* Product
  architecture; Floor; Walls; Roof;
  Finish plumbing (build
  subassemblies); Rough
  plumbing (build subassemblies);
  Window (build opening
  subassemblies); Material
  handling
Subcontractors, 6, 7, 8, 9, 10, 142
Supply chain management, 6–7, 10,
  29, 223
Synchronous flow, 35, 132–136
Systems Building Research Alliance
  benchmarking study, 12, 13, 14,
    18, 19, 20, 21, 22, 23, 39, 45
  home sales process study, 29
  lean production study, 32, 127

T

TAKT time, 35, 98, 99
Toyota Production System, 32
Touch-up, 87

U

V

VSM. *See* Value stream map
Value stream map, 38, 103–110
  use, 110–144
Variability. *See* Cycle time
  variation; Work content;
  Product choice
Visual management, 33, 201

W

Walls. *See* End walls; Marriage
  wall; Partition walls; Side wall;
  Rough electric (install in walls);
  Rough plumbing (install in
  walls and tubs); Sheath walls;
  Material handling
  (subassemblies)
Warehouse. *See* Material handling
  (building material)
Waste (seven types), 32
Whole-house homebuilding, 6
Window
  build opening subassemblies,
    51, 125
  install, 79–80
Wood frame construction, 1, 17, 97
Work content, 99
  estimation, 100
Workstation
  assigning activities to, 144–145
  arranging in layout, 144–165
  detailed design, 201–220
  linking with material handling
    systems. *See* Material
    handling
  vs. capacity, 145
Wrap for shipment
  prep, drop and wrap for
    shipment, 76–77
  final wrap and prep for
    shipment, 91

X,Y,Z

LaVergne, TN USA
11 April 2011
223800LV00007B/5/P

9 780983 321200